P9-BYR-782

HOME LINKS

Everyday
Mathematics®

The University of Chicago School Mathematics Project

Mc
Graw
Hill
Education

The University of Chicago School Mathematics Project

Max Bell, Director, *Everyday Mathematics* First Edition; James McBride, Director, *Everyday Mathematics* Second Edition; Andy Isaacs, Director, *Everyday Mathematics* Third, CCSS, and Fourth Editions; Amy Dillard, Associate Director, *Everyday Mathematics* Third Edition; Rachel Malpass McCall, Associate Director, *Everyday Mathematics* CCSS and Fourth Editions; Mary Ellen Dairyko, Associate Director, *Everyday Mathematics* Fourth Edition

Authors
Max Bell, Jean Bell, John Bretzlauf, Amy Dillard, Robert Hartfield, Andy Isaacs, James McBride, Cheryl G. Moran, Kathleen Pitvorec, Peter Saecker

Fourth Edition Grade 2 Team Leader
Cheryl G. Moran

Writers
Camille Bourisaw, Mary Ellen Dairyko, Gina Garza-Kling, Rebecca Williams Maxcy, Kathryn M. Rich

Open Response Team
Catherine R. Kelso, Leader; Steve Hinds

Differentiation Team
Ava Belisle-Chatterjee, Leader; Jean Marie Capper, Martin Gartzmann,

Digital Development Team
Carla Agard-Strickland, Leader; John Benson, Gregory Berns-Leone, Juan Camilo Acevedo

Virtual Learning Community
Meg Schleppenbach Bates, Cheryl G. Moran, Margaret Sharkey

Technical Art
Diana Barrie, Senior Artist; Cherry Inthalangsy

UCSMP Editorial
Don Reneau, Senior Editor; Rachel Jacobs, Kristen Pasmore, Luke Whalen

Field Test Coordination
Denise A. Porter

Field Test Teachers
Kristin Collins, Debbie Crowley, Brooke Fordice, Callie Huggins, Luke Larmee, Jaclyn McNamee, Vibha Sanghvi, Brook Triplett

Digital Field Test Teachers
Colleen Girard, Michelle Kutanovski, Gina Cipriani, Retonyar Ringold, Catherine Rollings, Julia Schacht, Christine Molina-Rebecca, Monica Diaz de Leon, Tiffany Barnes, Andrea Bonanno-Lersch, Debra Fields, Kellie Johnson, Elyse D'Andrea, Katie Fielden, Jamie Henry, Jill Parisi, Lauren Wolkhamer, Kenecia Moore, Julie Spaite, Sue White, Damaris Miles, Kelly Fitzgerald

Contributors
William B. Baker, John Benson, Jeanna Mills DiDomenico, James Flanders, Lila K. S. Goldstein, Funda Gönülateş, Allison Greer, Lorraine M. Males, John P. Smith III, Kathleen Clark, Patti Satz, Penny Williams

Center for Elementary Mathematics and Science Education Administration
Martin Gartzman, Executive Director; Jose J. Fragoso, Jr., Meri B. Forhan, Regina Littleton, Laurie K. Thrasher

External Reviewers
The *Everyday Mathematics* authors gratefully acknowledge the work of the many scholars and teachers who reviewed plans for this edition. All decisions regarding the content and pedagogy of *Everyday Mathematics* were made by the authors and do not necessarily reflect the views of those listed below.

Elizabeth Babcock, California Academy of Sciences; Arthur J. Baroody, University of Illinois at Urbana-Champaign and University of Denver; Dawn Berk, University of Delaware; Diane J. Briars, Pittsburgh, Pennsylvania; Kathryn B. Chval, University of Missouri–Columbia; Kathleen Cramer, University of Minnesota; Ethan Danahy, Tufts University; Tom de Boor, Grunwald Associates; Louis V. DiBello, University of Illinois at Chicago; Corey Drake, Michigan State University; David Foster, Silicon Valley Mathematics Initiative; Funda Gönülateş, Michigan State University; M. Kathleen Heid, Pennsylvania State University; Natalie Jakucyn, Glenbrook South High School, Glenview, IL; Richard G. Kron, University of Chicago; Richard Lehrer, Vanderbilt University; Susan C. Levine, University of Chicago; Lorraine M. Males, University of Nebraska-Lincoln; Dr. George Mehler, Temple University and Central Bucks School District, Pennsylvania; Kenny Huy Nguyen, North Carolina State University; Mark Oreglia, University of Chicago; Sandra Overcash, Virginia Beach City Public Schools, Virginia; Raedy M. Ping, University of Chicago; Kevin L. Polk, Aveniros LLC; Sarah R. Powell, University of Texas at Austin; Janine T. Remillard, University of Pennsylvania; John P. Smith III, Michigan State University; Mary Kay Stein, University of Pittsburgh; Dale Truding, Arlington Heights District 25, Arlington Heights, Illinois; Judith S. Zawojewski, Illinois Institute of Technology

Note
Too many people have contributed to earlier editions of *Everyday Mathematics* to be listed here. Title and copyright pages for earlier editions can be found at http://everydaymath.uchicago.edu/about/ucsmp-cemse/.

www.everydaymath.com

Copyright © McGraw-Hill Education

All rights reserved. The contents, or parts thereof, may be reproduced in print form for non-profit educational use with *Everyday Mathematics,* provided such reproductions bear copyright notice, but may not be reproduced in any form for any other purpose without the prior written consent of McGraw-Hill Education, including, but not limited to, network storage or transmission, or broadcast for distance learning.

Send all inquiries to:
McGraw-Hill Education
STEM Learning Solutions Center
8787 Orion Place
Columbus, OH 43240

ISBN: 978-0-02-137959-0
MHID: 0-02-137959-9

Printed in the United States of America.

13 LHS 23

Contents

Unit 9

Unit 1: Family Letter

Introduction to *Second Grade Everyday Mathematics*

Welcome to *Second Grade Everyday Mathematics,* which is part of an elementary school mathematics curriculum developed by the University of Chicago School Mathematics Project (UCSMP).

Here we describe several features of the program to familiarize you with the structure of *Everyday Mathematics* and the expectations we have for children.

A Problem-Solving Approach Based on Everyday Situations By connecting what children learn to their experiences both in and out of school, *Everyday Mathematics* presents basic math skills and concepts in meaningful contexts so that the mathematics becomes "real."

Frequent Practice of Basic Skills In *Everyday Mathematics,* children practice basic skills in a number of different ways—but *not* through tedious drilling. Second graders complete daily review exercises covering a host of topics. They learn to find patterns on the number line and the number grid, explore addition and subtraction fact families in a variety of formats, work with Quick Looks and ten frames, and play games specifically designed to help them develop basic skills.

An Instructional Approach That Revisits Concepts Regularly The best way for children to develop their mathematical understanding is to regularly revisit skills and concepts they encountered earlier. Rather than presenting mathematics as isolated bits of content, the *Everyday Mathematics* curriculum is designed to build on children's learning throughout the year. Research shows that repeated exposure to math concepts and skills over time develops children's abilities to recall knowledge from long-term memory.

A Curriculum That Explores Mathematical Content and Practices
The rich problem-solving environment provided by *Everyday Mathematics* helps children develop critical-thinking skills. They learn to solve new kinds of problems, explain their thinking to others, and make sense of other children's thinking.

Second Grade Everyday Mathematics emphasizes the following content:

Numbers and Operations in Base 10
Understanding place value through counting, making coin exchanges, reading and writing numbers, and comparing numbers; using place-value understanding to add and subtract whole numbers.

Operations and Algebraic Thinking Solving addition and subtraction problems; developing fluency with addition and subtraction facts; exploring fact families (related addition and subtraction facts, such as $2 + 5 = 7$, $5 + 2 = 7$, $7 - 5 = 2$, and $7 - 2 = 5$); gaining foundations for multiplication and division.

Measurement and Data Estimating lengths and using tools to measure length; telling time to the nearest 5 minutes; solving problems involving money; collecting, organizing, and representing data with tables and graphs.

Geometry Recognizing and drawing 2-dimensional shapes and identifying select 3-dimensional shapes.

Everyday Mathematics provides you with many opportunities to share in your child's mathematical experience and monitor the progress your child makes. Throughout the year you will receive Family Letters like this one to keep you informed of the mathematical content your child is studying in each unit. Each letter includes a vocabulary list, suggested Do-Anytime Activities for you and your child, and an answer guide to selected Home Link (homework) activities. You will enjoy seeing your child's confidence and comprehension soar as he or she connects mathematics to everyday life.

We look forward to an exciting year!

Unit 1: Establishing Routines

This unit reviews and extends mathematical concepts that were developed in *First Grade Everyday Mathematics*. In Unit 1 children will do the following:

- Use number lines to count, compare numbers, add, and subtract.

- Count in several different intervals, such as up by 2s, up by 10s, back by 10s from 100.

- Review whole numbers by completing assigned tasks, such as writing the number that comes after 509, writing the number that comes before 1,001, and writing the number word for 50.

- Count coins and find the values of coin combinations.

- Work with a number grid to reinforce place-value skills and observe number patterns.

-9	-8	-7	-6	-5	-4	-3	-2	-1	0
1	2	3	4	5	6	7	8	9	10
11	12	13	14	15	16	17	18	19	20
21	22	23	24	25	26	27	28	29	30

- Review equivalent names for numbers, which are different ways numbers can be expressed. For example, some equivalent names for 10 are $5 + 5$, $20 - 10$, ten, and ~~HHT~~ ~~HHT~~.

- Play games, such as *Fishing for 10,* to strengthen number skills and develop fact fluency.

- Explore patterns involving odd and even numbers.

- Review and use the symbols $>$ (is greater than), $<$ (is less than), and $=$ (is equal to).

Do-Anytime Activities

Try these interesting and rewarding activities to practice concepts taught in this unit:

- Discuss examples of mathematics in everyday life: times in TV listings, distances or speeds on road signs, prices in ads or store displays, recipe measurements, and so on.

- Discuss the rules for working with a partner or in a group.

  - Be polite.
  - Help each other.
  - Share.
  - Listen to your partner.
  - Respond to your partner.
  - Talk about problems.
  - Take turns.
  - Speak quietly.

- Discuss household tools that can be used to help solve mathematical problems, such as tape measures, thermometers, and clocks.

- Count combinations of pennies, nickels, dimes, and quarters.
- Look for number lines on everyday objects, such as rulers, speedometers, and thermometers.

Vocabulary   Important terms in Unit 1:

Math Message A daily activity that children complete independently, usually as a lead-in to the day's lesson. Example: *Make tally marks to show how many children are here today.*

Math Journal A book used by each child. It contains examples, instructions, and problems, as well as space to record answers and observations.

toolkits Individual zippered bags or boxes used in the classroom. Each toolkit contains various items—such as a ruler, play money, and number cards—that are used to help children understand mathematical ideas.

Mental Math and Fluency A daily whole-class oral or written activity, often emphasizing computation children learn to do in their heads.

number grid A table in which numbers are arranged consecutively, usually in rows of 10. A move from one number to the next left or right in a row is a change of 1; a move from one number to the next up or down in a column is a change of 10.

Exploration A small-group, hands-on activity designed to introduce or extend a mathematical topic.

Math Boxes Math problems in the *Math Journal* that provide opportunities for children to review and practice previously introduced skills.

Home Links *Everyday Mathematics* daily homework. Each Home Link includes problems and activities intended for follow-up and enrichment at home.

As You Help Your Child with Homework

When your child brings home an assignment, you may want to go over the instructions together, clarifying them as necessary. Each Family Letter will contain answers, such as the following, to guide you through the unit's Home Links.

Lesson 1-11
1. Answers vary.
2. Answers vary.
3. Answers vary.
4. <
5. >
6. >

7. =
8. Answers vary.
9. Answers vary.

Lesson 1-12
158; Answers vary.

Family Letter

Open Response and Reengagement Lessons

A two-day lesson in each unit of *Second Grade Everyday Mathematics* is an Open Response and Reengagement lesson. In these lessons children solve interesting problems using their own strategies and reasoning. On Day 1 children solve an open response problem—a problem with more than one possible strategy or solution. On Day 2 the class discusses children's work from Day 1 to "reengage" with the problem and learn more about the mathematics involved. Children then revise their work based on what they learn from the discussion.

These lessons are not assessments, but opportunities for children to solve approachable problems that require persistence. Children's work on Day 1 reveals both strengths and weaknesses, allowing the discussion on Day 2 to focus on areas that need improvement. From these discussions, children find that learning from mistakes is a natural part of mathematical problem solving. Explaining their thinking and listening to the explanations of others builds children's confidence. At the same time, children see that there is more than one way to solve a problem, which promotes creative solutions to new problems. Having an opportunity to revise their work helps children realize that they can be successful tackling hard tasks if they think about them and keep trying.

The open response problem in this unit asks children to look for patterns in a number grid and use the patterns to identify missing numbers in a "number-grid puzzle." They also write explanations about how they figured out two of the missing numbers.

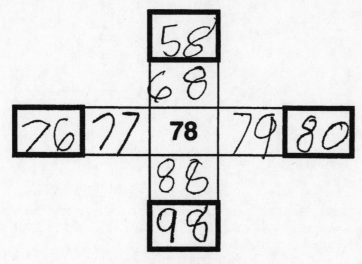

Number-Grid Puzzle

These lessons continue work on problem solving that is central to *Everyday Mathematics* across all the grades. We hope you enjoy seeing your child become a confident problem solver. Ask your child to talk to you about the problems and his or her mathematical thinking throughout the year.

Relations: <, >, =

Family Note

In *Second Grade Everyday Mathematics,* children "do mathematics." We expect that children will want to share their enthusiasm for the mathematics activities they do in school with their families. Your child will bring home assignments and activities to do as homework throughout the year. These assignments, called Home Links, will be identified by the house at the top right corner of this page. The assignments will not take very much time to complete, but most of them involve interaction with an adult or an older child.

There are many reasons for including Home Links in the second-grade program:

- The assignments encourage children to take initiative and responsibility for completing them. As you respond with encouragement and assistance, you help your child build independence and self-confidence.

- Home Links reinforce newly learned skills and concepts. They provide opportunities for children to think and practice at their own pace.

- These assignments are often designed to relate what is done in school to children's lives outside school. This helps tie mathematics to the real world, which is very important in the *Everyday Mathematics* program.

- The Home Links will give you a better idea of the mathematics your child is learning in school.

Generally, you can help by listening and responding to your child's requests and comments about mathematics. You also can help by linking numbers to real life, pointing out ways in which you use numbers (time, TV channels, page numbers, telephone numbers, bus routes, shopping lists, and so on). Extending the notion that "children who are read to, read," *Everyday Mathematics* supports the belief that children who have someone do math with them will learn mathematics. Playful counting and thinking games that are fun for both you and your child are very helpful for such learning.

Please return the second page of this Home Link to school tomorrow.

Relations: <, >, =
(continued)

Family Note

MRB

Page 75

This icon will often appear on the Home Links. This icon tells children where to look in *My Reference Book* to find more information about the concept or skill addressed in the Home Link. In today's lesson, we reviewed and practiced using the <, >, and = symbols. For information about relation symbols, see page 75 in *My Reference Book*.

Show someone at home your *My Reference Book*. Together find three things you found interesting and write them below.

① _____

② _____

③ _____

Explain to someone at home how to do Problems 4–7.
Then write <, >, or = on each blank. Use *My Reference Book* to look up the symbols.

④ 8 _____ 12

⑤ 25¢ _____ 18¢

⑥ 103 _____ 53

⑦ 79¢ _____ 79¢

Write numbers in the blanks to make up your own.

⑧ _____ < _____

⑨ _____ > _____

Base-10 "Buildings"

Family Note

In today's lesson, children were introduced to Grade 2 Explorations. Explorations are small-group, hands-on activities designed to introduce or extend mathematical topics. One of the Explorations reviewed the values represented by base-10 blocks. Children made "buildings" with the blocks and calculated the values of the buildings.

Please return this Home Link to school tomorrow.

Look at the picture of a "building" that is made with base-10 blocks.

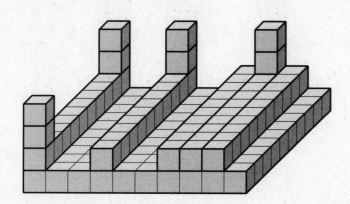

▪	=	1
I	=	10
□	=	100

Count the value of the flats, longs, and cubes that make up the building. What number does the building show? Use the symbols in the box to help you. _____

Write about how you counted the blocks.

_____

_____

_____

Fact Strategies

Unit 2 focuses on developing strategies for solving addition facts. In *Everyday Mathematics* children learn basic facts by first focusing on specific groups of facts that can be solved using a particular strategy. Children build fluency and automatic recall as they develop strategies for all the different groups of facts. Achieving automatic recall of basic addition facts will enable your child to solve multidigit computation problems with ease later in the year.

Everyday Mathematics Program Routines

Your child will use two new program routines in this unit. **Name-collection boxes** provide a space for children to collect equivalent names for numbers. **Frames-and-Arrows diagrams** show sequences of numbers following a certain pattern. More information about these routines can be found in the Family Notes on Home Links 2-10 and 2-12.

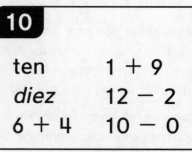

A name-collection box

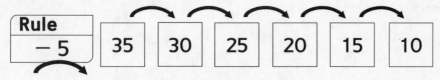

A frames-and-arrows diagram

Please keep this Family Letter for reference as your child works through Unit 2.

Vocabulary   Important terms in Unit 2:

label A unit, descriptive word, or phrase used to put a number or numbers in context. Using a label reinforces the idea that numbers often refer to something.

unit box A box that contains the label or unit of measure for the numbers in a problem. *For example:* In number stories that involve counting children in the class, the word *children* would go in the unit box.

Unit
children

number model A number sentence or other representation that fits a number story or situation. *For example:* $5 + 8 = 13$ models the number story "There are 5 children skating. There are 8 children playing ball. How many children are there in all?"

number story A story involving numbers that is made up by children, teachers, or parents. Children solve problems posed in number stories using many different methods. In Grade 2, number stories focus on addition and subtraction.

doubles fact An addition fact in which a number is added to itself, such as $4 + 4 = 8$ and $9 + 9 = 18$.

combination of 10 An addition fact with a sum of 10, such as 6 + 4 = 10 and 7 + 3 = 10.

addend Any one of a set of numbers that is added. *For example:* In 5 + 3 = 8, the addends are 5 and 3.

turn-around rule for addition A rule that says you can add two numbers in either order and get the same result (for example, 3 + 5 = 8 and 5 + 3 = 8).

name-collection box An empty box used to collect equivalent names for a given number. The tag in the top left corner identifies the number whose names are collected in the box.

Frames-and-Arrows diagram A diagram used to represent a number sequence, which is a list of numbers that follow some rule. A Frames-and-Arrows diagram consists of frames connected by arrows that show the path from one frame to the next. Each arrow represents a rule that determines which number goes in the next frame so that all of the frames contain the numbers in the sequence.

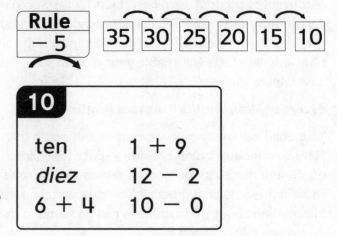

Building Skills through Games

In Unit 2 your child will explore place-value concepts and practice addition facts by playing the following games.

The Exchange Game
Each player rolls a die and collects that number of $1 bills from the bank. As players accumulate bills, they exchange ten $1 bills for one $10 bill and ten $10 bills for one $100 bill.

Evens and Odds
Each player draws a card. If the card shows an even number, the player writes it as a sum of two equal addends. If the card shows an odd number, the player writes it as the sum of two equal addends plus or minus 1. *For example:* A player who draws a 6 writes 3 + 3 = 6, and a player who draws a 7 writes 3 + 3 + 1 = 7 or 4 + 4 − 1 = 7.

Name That Number
Players turn over a card to show a target number that must be renamed using any combination of five faceup cards.

| 4 | 10 | 8 | 12 | 2 | | 6 |

$$6 = 8 - 2$$
$$6 = 10 - 4$$
$$6 = 4 + 2$$

Do-Anytime Activities

To work with your child on the mathematical concepts taught up to this point in the school year, try these interesting and rewarding activities:

1. Talk with your child about why it is important to learn basic facts.

2. Create addition number stories about common objects in your child's environment.

3. Have your child explain his or her favorite fact strategy to you.

4. Name a number and ask your child to think of several different ways to represent it. *For example:* 10 can be represented as $1 + 9$, ten tally marks, the word *ten,* and so on.

5. Ask your child to make fair exchanges between $1 and $10 bills or among coins.

6. Call out numbers and ask your child whether the numbers are even or odd.

As You Help Your Child with Homework

Your child will regularly bring home assignments with instructions you may want to go through together, clarifying them as necessary. The following represent the answers to every problem in the Unit 2 Home Links.

Home Link 2-1
1. 1
2. 100
3. 10
4. Sample answers: 5; 50
5. $14
6. $29
7. $120

Home Link 2-2
1. Answers vary.
2. 4
3. 8
4. 10
5. 14

Home Link 2-3
1. **a.** 4 **b.** 10 **c.** 0 **d.** 2 **e.** 14 **f.** 3 **g.** 16
2. Sample answers: $10 + 0$; $9 + 1$; $8 + 2$; $7 + 3$; $6 + 4$; $5 + 5$

Home Link 2-4
1. $10 + 0 = 10$; $9 + 1 = 10$; $8 + 2 = 10$; $7 + 3 = 10$; $6 + 4 = 10$; $5 + 5 = 10$; $4 + 6 = 10$; $3 + 7 = 10$; $2 + 8 = 10$; $1 + 9 = 10$; $0 + 10 = 10$
2. 11; $8 + 2 = 10$
3. 11; $4 + 6 = 10$
4. 12; $9 + 1 = 10$

Home Link 2-5
1. Answers: 9, 11; Helper fact: 10
2. Answers: 13, 15; Helper fact: 14
3. 7; Helper fact: $4 + 4 = 8$ or $3 + 3 = 6$

Home Link 2-6

1. $2 + 4 = 6$; $4 + 2 = 6$

2. $3 + 5 = 8$; $5 + 3 = 8$

3. $4 + 6 = 10$; $6 + 4 = 10$

4. $3 + 8 = 11$; $8 + 3 = 11$

5. 10   6. 10

7. 10   8. 10

Home Link 2-7

1. $6 + 8 = 14$; $8 + 6 = 14$

2. 15; $3 + 12 = 15$

   3; $3 + 8 = 11$

3. **a.** 3   **b.** 5   **c.** 7   **d.** 6

Home Link 2-8

1. Answer: 66

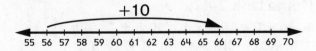

2. Answer: 66

1	2	3	4	5	6	7	8	9	10
11	12	13	14	15	16	17	18	19	20
21	22	23	24	25	26	27	28	29	30
31	32	33	34	35	36	37	38	39	40
41	42	43	44	45	46	47	48	49	50
51	52	53	54	55	56	57	58	59	60
61	62	63	64	65	66	67	68	69	70
71	72	73	74	75	76	77	78	79	80

Home Link 2-9

Children should circle 6, 18, 8, 14, 4, 10, 16, 2, 20, and 12; they should underline 9, 3, 11, 17, 15, 1, 7, 19, 13, and 5.

$7 \rightarrow 3 + 3 + 1$

$8 \rightarrow 4 + 4$

$11 \rightarrow 6 + 6 - 1$

$14 \rightarrow 7 + 7$

$17 \rightarrow 8 + 8 + 1$

$10 \rightarrow 5 + 5$

Home Link 2-10

1. Answers vary.

2. Sample answers: Ten, $11 - 1$, $10 - 0$, $10 + 0$, $5 + 5$, $13 - 3$, $8 + 1 + 1$, $2 + 2 + 2 + 2 + 2$,

   ~~||||~~ ~~||||~~ ,

   X X X X X
   X X X X X

3. Answers vary.

Home Link 2-11

1. Sample answers: $6 + 6 = 12$; $10 + 2 = 12$

2. Sample answers: $9 - 4 = 5$; $6 - 1 = 5$

3. Sample answers: $9 - 3 = 6$; $4 + 2 = 6$

4. 3   5. 1

6. 8   7. 10

Home Link 2-12

1. Rule + 2   32   34   36   38   40

2. Rule − 5   45   40   35   30   25

3. Rule + 10   38   48   58   68   78

4. Rule + 3   8   11   14   17   20

5. Answers vary.

Money Exchanges

Family Note

In today's lesson we counted by 100s, 10s, and 1s to calculate the values of bill combinations. We also played *The Exchange Game* with money to practice making exchanges between $1, $10, and $100 bills.

Please return this Home Link to school tomorrow.

Answer the questions.

(1) How many $10 bills are the same as ten $1 bills? _____

(2) How many $1 bills are the same as one $100 bill? _____

(3) How many $10 bills are the same as one $100 bill? _____

Do your own.

(4) How many $1 bills are the same as _____ $10 bills? _____

For each problem, find how much money there is in all.

Example:    = $23

(5)    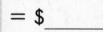 = $_____

(6)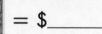
      = $_____

(7)   = $_____

McGraw-Hill Education/Studio/nfio, Michael Houghton

Family Note

Today we continued to explore ways to analyze the value of our combinations. We also learned to exchange bills with money to make mathematical exchanges, such as ten $1's for one $10 bill.

Please return the phone this message tomorrow.

Answer the questions.

1. How many $10 bills are the same as ten $1 bills? _____

2. How many $1 bills are the same or one $100 bill? _____

3. How many $10 bills are the same as one $100 bill? _____

Do your own.

4. How many $1 bills are the same as _____ $10 bills?

For each problem, find how much money there is in all.

Example

Writing Addition Number Stories

Family Note

Before beginning this Home Link, review the vocabulary from the Unit 2 Family Letter with your child: *number story, label, unit box,* and *number model.* Encourage your child to make up and solve addition number stories and write addition number models for the stories. Stress that the answer to a question makes more sense if it has a label.

Please return this Home Link to school tomorrow.

① Tell someone at home what you know about number stories, labels, unit boxes, and number models. Write an addition number story for the picture. Write the answer and an addition number model.

MRB
24-29

Unit

lions

Story: _____

_____

_____

_____

_____

_____

Number model:

_____

Practice

② 2 + 2 = _____

③ 4 + 4 = _____

④   5
   + 5
     ———

⑤   7
   + 7
     ———

Doubles Facts and Combinations of 10

Family Note

In second grade children learn and practice many strategies to help them develop fluency with basic addition facts. These strategies are based on facts that children studied in first grade: doubles facts (facts in which a number is added to itself) and combinations of 10 (number pairs that add to 10). Today we sorted facts based on whether they were doubles, combinations of 10, or both.

Please return this Home Link to school tomorrow.

---

① Complete the addition facts.

MRB
40-41

a. $2 + 2 =$ _____

b. _____ $= 5 + 5$

c. _____ $= 0 + 0$

d. $1 + 1 =$ _____

Unit
birds

e.
$$\begin{array}{r} 7 \\ + 7 \\ \hline \end{array}$$

f.
$$\begin{array}{r} 3 \\ + \\ \hline 6 \end{array}$$

g.
$$\begin{array}{r} 8 \\ + 8 \\ \hline \end{array}$$

② Write four different addition facts with 10 as the sum.

Example: $10 = \underline{\quad 4 \quad} + \underline{\quad 6 \quad}$

a. $10 =$ _____ $+$ _____

b. $10 =$ _____ $+$ _____

c. _____ $+$ _____ $= 10$

d. _____ $+$ _____ $= 10$

The Making-10 Strategy

Family Note

Children learned a new strategy called "the making-10 strategy" to help them develop fluency with the basic addition facts. Success with this strategy depends on children knowing the number pairs that add to 10 or the basic addition facts that have a sum of 10.

Please return this Home Link to school tomorrow.

① Write all the addition facts that have a sum of 10.
Hint: There are 11 different facts.

Use combinations of 10 to help figure out the sums.

Example: 6 + 5 = __*11*__

  Helper combination of 10: __*6*__ + __*4*__ = __*10*__

② 8 + 3 = _____
Helper combination of 10: _____ + _____ = _____

③ 4 + 7 = _____
Helper combination of 10: _____ + _____ = _____

④ 9 + 3 = _____
Helper combination of 10: _____ + _____ = _____

Tell someone at home how knowing combinations of 10 can help you solve other facts.

Helper Facts

Family Note

Today we learned about helper facts. Helper facts are facts that we already know and that we can use to help us find answers to facts we may not know. Because we learned the doubles facts (such as $2 + 2 = 4$ and $3 + 3 = 6$) in first grade, we can use them now as helper facts. For example, knowing $4 + 4 = 8$ can help us figure out the answer to $4 + 5$. We see that $4 + 5$ is 1 more than $4 + 4$, so the answer is 9. We can also use $4 + 4 = 8$ to figure out the answer to $4 + 3$. We see that $4 + 3$ is 1 less than $4 + 4$, so the answer is 7.

Please return this Home Link to school tomorrow.

Helper facts can help you figure out answers to other facts. Doubles facts can be helper facts.

Example:

$3 + 2 = ?$ $3 + 4 = ?$

Helper fact: $3 + 3 = 6$

$3 + 2 = \underline{\ \ 5\ \ }$ $3 + 4 = \underline{\ \ 7\ \ }$

MRB
42

(1) $5 + 4 = ?$ $5 + 6 = ?$

Helper fact: $5 + 5 = \underline{\hspace{1.5cm}}$

$5 + 4 = \underline{\hspace{1.5cm}}$ $5 + 6 = \underline{\hspace{1.5cm}}$

Unit
birds

(2) $? = 7 + 6$ $? = 7 + 8$

Helper fact: $\underline{\hspace{1.5cm}} = 7 + 7$

$\underline{\hspace{1.5cm}} = 7 + 6$ $\underline{\hspace{1.5cm}} = 7 + 8$

Solve. Write a helper fact that can help you figure out the answer.

(3) $4 + 3 = ?$

Helper fact: $\underline{\hspace{1.5cm}} + \underline{\hspace{1.5cm}} = \underline{\hspace{1.5cm}}$

$4 + 3 = \underline{\hspace{1.5cm}}$

Tell someone at home what you know about helper facts.

The Turn-Around Rule for Addition

Family Note

Today we learned about the turn-around rule for addition, which says that you can add two numbers in either order and get the same result. *For example:* 4 + 3 = 7 and 3 + 4 = 7. Knowing this rule can help children learn basic addition facts.

Please return this Home Link to school tomorrow.

Use the turn-around rule to write two different addition facts for each domino.

①  _____ + _____ = _____

_____ + _____ = _____

②  _____ + _____ = _____

_____ + _____ = _____

③  ④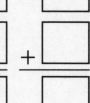

Tell someone at home what you know about the turn-around rule.

Facts Practice

Solve.

⑤   8
   + 2

⑥   2
   + 8

⑦   7
   + 3

⑧   3
   + 7

Unit

children

Copyright © McGraw-Hill Education. Permission is granted to reproduce for classroom use.

The Turn-Around Rule

Family Note

Today we continued exploring the turn-around rule. In this lesson, we created subtraction number stories and examined whether the turn-around rule works for subtraction. We decided the turn-around rule does NOT work for subtraction because changing the order of the numbers produces different results. Creating number stories, writing number models for number stories, and using the turn-around rule appropriately will all be revisited throughout the year. The problems on this page provide practice with the turn-around rule for addition.

Please return this Home Link to school tomorrow.

① Write two number models you can use to solve this number story.

Bill picked 6 peaches from a tree. Roberta picked 8. How many peaches do they have in all?

Number model: _____

Number model: _____

② Complete each number sentence and then write the turn-around number sentence.

Number sentence   Turn-around number sentence

12 + 3 = _____   _____

8 + _____ = 11   _____

Practice

③ Solve the facts.

a. 3 + _____ = 6   b. _____ + 5 = 10

c. 14 = _____ + 7   d. 12 = 6 + _____

Counting Up

Family Note

Everyday Mathematics encourages children to use a variety of strategies to solve problems. This allows children to develop number sense rather than simply memorizing steps or learning a shortcut. In today's lesson children used various strategies to add, including counting up on a number line and on a number grid.

Please return this Home Link to school tomorrow.

Find the sum in two different ways.

$$56 + 10 = ?$$

(1) Use the number line and show your hops. Record your answer.

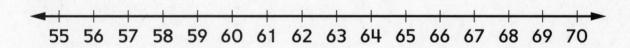

55 56 57 58 59 60 61 62 63 64 65 66 67 68 69 70

Answer: 56 + 10 = _____

(2) Use a number grid and draw arrows to show your counts. Record your answer.

1	2	3	4	5	6	7	8	9	10
11	12	13	14	15	16	17	18	19	20
21	22	23	24	25	26	27	28	29	30
31	32	33	34	35	36	37	38	39	40
41	42	43	44	45	46	47	48	49	50
51	52	53	54	55	56	57	58	59	60
61	62	63	64	65	66	67	68	69	70
71	72	73	74	75	76	77	78	79	80

Answer: 56 + 10 = _____

Even Numbers and Equal Addends

Family Note

Today we identified numbers as even or odd. We learned that when we count by 2s, we can look for a pattern in the digits in the ones place to help us identify even and odd numbers. We wrote number models to express even numbers as the sums of two equal addends and odd numbers as the sums of two equal addends plus or minus 1. We also learned a new game called *Evens and Odds*.

Please return this Home Link to school tomorrow.

Draw a circle around the even numbers.
Draw a line below the odd numbers.

6   9   18   8   3   14   11   4   17   10

15   1   7   19   16   13   2   20   5   12

Pick one of your circled numbers. Tell someone at home how you know it is even.

Pick one of your underlined numbers. Tell someone at home how you know it is odd.

Match.

7	4 + 4
8	8 + 8 + 1
11	6 + 6 − 1
14	3 + 3 + 1
17	7 + 7
10	5 + 5

Name-Collection Boxes

Family Note

Beginning in *First Grade Everyday Mathematics*, children use name-collection boxes to help them collect equivalent names for the same number. These boxes help children appreciate the idea that numbers can be expressed in many different ways.

A name-collection box is an open box with a tag in the corner. The tag identifies the number whose names are collected in the box. In second grade typical names include sums, differences, tally marks, number words, and arrays. At higher grades, names may include products, quotients, and the results of several mathematical operations.

Encourage your child to name a number in different ways—for example, use tally marks, write addition and subtraction problems, or draw pictures of objects. Some name-collection activities are shown below.

10 ← Tag for box

Name-collection box

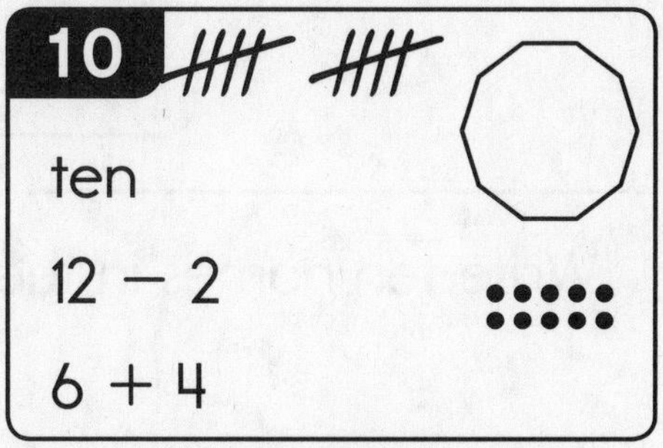

10 $\cancel{||||}$ $\cancel{||||}$

ten

$12 - 2$

$6 + 4$

Items in the name-collection box above represent the number 10. Some names contain numbers, and some do not.

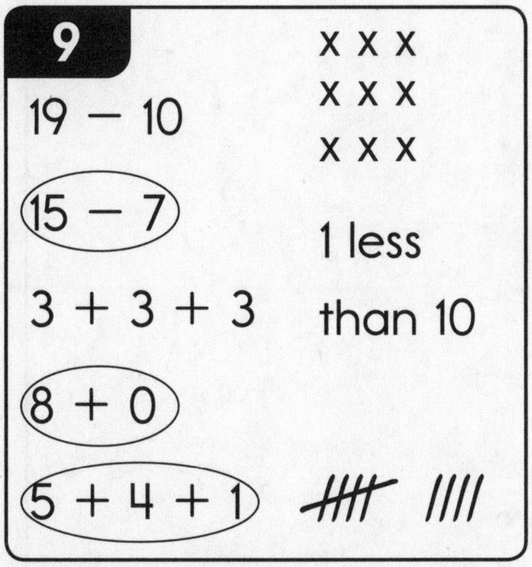

9

$19 - 10$

$\boxed{15 - 7}$

$3 + 3 + 3$

$\boxed{8 + 0}$

$\boxed{5 + 4 + 1}$

x x x
x x x
x x x

1 less
than 10

$\cancel{||||}$ ||||

Sometimes children must circle names that do not belong in the box.

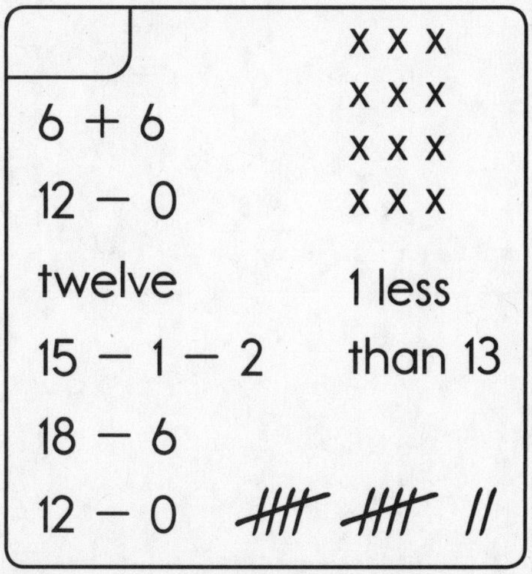

$6 + 6$

$12 - 0$

twelve

$15 - 1 - 2$

$18 - 6$

$12 - 0$

x x x
x x x
x x x
x x x

1 less
than 13

$\cancel{||||}$ $\cancel{||||}$ ||

Sometimes children must fill in the tag for the numbers shown in the box. The tag here should read 12.

Please return the second page of this Home Link to school tomorrow.

Name-Collection Boxes (continued)

① Give the Family Note to someone at home.
Show that person the name-collection box below.
Explain what a name-collection box is used for.

8		
2 + 6	4 + 4	x x x x
eight	12 − 4	x x x x
ocho	10 − 2	8 − 0
8 + 0	3 + 5	⾞卌 ///

② Write ten names in the 10 box.

③ Make up your own name-collection box.
Write at least ten names in the box.

10

Playing *Name That Number*

Family Note

In today's lesson children discussed how to name a number in different ways. We played a game called *Name That Number* by using addition and subtraction to name a target number.

Please return this Home Link to school tomorrow.

Write two number sentences to show each target number.

Example:

_____ 6 + 2 = 8 _____   _____ 4 + 4 = 8 _____

①

_____   _____

②

_____   _____

③ 9 2 3 5 4 ▨ 6

_____   _____

Practice

Solve.

Unit
leaves

④ 10 = 7 + _____   ⑤ 10 = _____ + 9

⑥ 10 = 2 + _____   ⑦ 10 = 0 + _____

Frames and Arrows

Family Note

Today your child used Frames-and-Arrows diagrams. These diagrams show sequences of numbers in which one number follows another according to a rule. Frames-and-Arrows diagrams are made up of shapes called *frames* and arrows connecting the frames. Each frame contains one of the numbers in the sequence. Each arrow stands for the rule, which tells how to find the number that goes in the next frame. Here is an example of a Frames-and-Arrows diagram. The arrow rule is "Add 2."

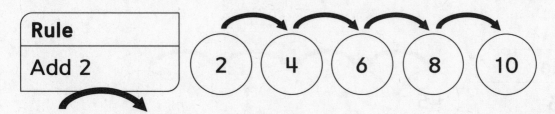

In a Frames-and-Arrows problem, some of the information is left out. To solve the problem, you have to find the missing information. Here are two examples of Frames-and-Arrows problems:

Example 1: Fill in the empty frames according to the rule.

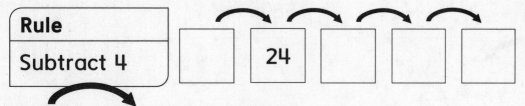

Solution: Write 28, 20, 16, and 12 in the empty frames.

Example 2: Write the arrow rule in the empty box.

Solution: The arrow rule is Add 5 or + 5.

Ask your child to tell you about Frames-and-Arrows diagrams. Take turns with your child making up and solving Frames-and-Arrows problems like the examples given above.

Please return the second page of this Home Link to school tomorrow.

Frames and Arrows (continued)

Tell someone at home what you know about Frames-and-Arrows problems. Fill in the empty frames and rule boxes.

①
Rule

+ 2
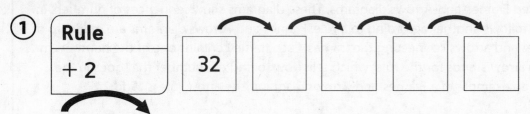

②
Rule

− 5
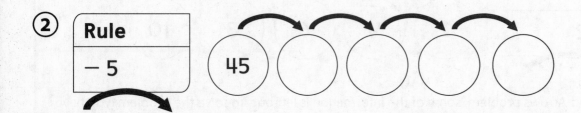

③
Rule

+ 10
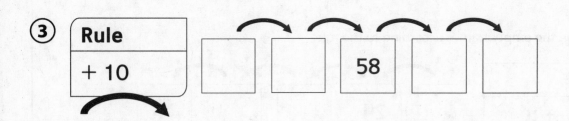

④
Rule

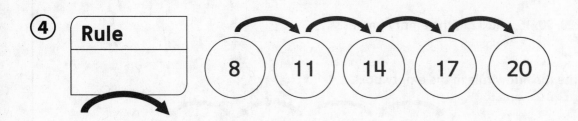

⑤ Do your own.

Rule

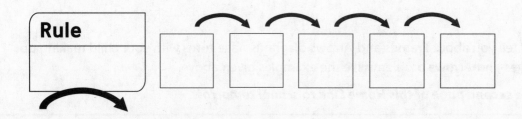

38   thirty-eight

More Fact Strategies

In Unit 3 your child will explore additional strategies for solving basic facts, focusing on strategies for solving subtraction facts. Children solve subtraction number stories and practice facts using games and routines.

In *Everyday Mathematics* children learn several strategies for solving subtraction facts. By becoming familiar with a variety of strategies, children have the opportunity to choose a strategy that works best to solve a particular fact. The goal is not for every child to master every strategy; the goal is for children to find the strategies they best understand and can most successfully apply. By encouraging discovery and practice, working with multiple strategies helps children develop fluency with subtraction facts, which will be important for computation with multidigit numbers later in the year.

Math Tools

Your child will use **Fact Triangles,** the *Everyday Mathematics* version of flash cards, to practice and review addition and subtraction facts. Each Fact Triangle shows related addition and subtraction facts made from the same three numbers, which helps your child understand the relationships among the facts. Home Link 3-3 provides a more detailed description of Fact Triangles and includes a set of Fact Triangles that your child can use to practice addition and subtraction facts at home.

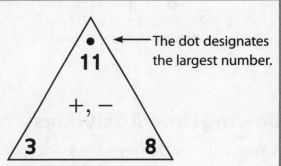

The dot designates the largest number.

A Fact Triangle showing the fact family for 3, 8, and 11

Vocabulary   Important terms in Unit 3:

related facts Addition and subtraction facts that use the same three numbers. For example, $2 + 3 = 5$ is related to $5 - 2 = 3$, and $9 + 8 = 17$ is related to $8 + 9 = 17$. All the facts in a **fact family** are related facts.

addition/subtraction fact family A collection of related addition and subtraction facts involving the same numbers. Most addition and subtraction fact families include two addition and two subtraction facts. For example, the addition/subtraction fact family for the numbers 2, 4, and 6 consists of the following:

$$2 + 4 = 6 \qquad 4 + 2 = 6$$
$$6 - 4 = 2 \qquad 6 - 2 = 4$$

Fact families involving doubles facts consist of only two facts. For example, the addition/subtraction fact family for the numbers 7, 7, and 14 consists of the following:

$$7 + 7 = 14 \qquad 14 - 7 = 7$$

− 0 facts Subtraction facts in which the number 0 is subtracted from another number, such as $7 - 0 = 7$ and $10 - 0 = 10$.

– 1 facts Subtraction facts in which the number 1 is subtracted from another number, such as 9 – 1 = 8 and 6 – 1 = 5.

"What's My Rule?" problem A problem in which number pairs are related to each other according to a rule or rules. A rule can be represented by a **function machine.**

in	out
3	8
5	10
8	13

"What's My Rule?" table

function machine In *Everyday Mathematics,* an imaginary device that receives input numbers and produces output numbers according to a set rule.

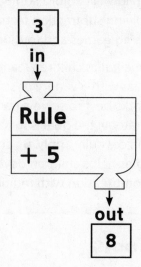

Do-Anytime Activities

To work with your child on the concepts taught in this and previous units, try these interesting and rewarding activities:

1. Talk with your child about why it is important to learn basic facts.

2. Create addition and subtraction stories about everyday subjects.

3. Have your child explain a favorite fact strategy to you.

4. Name pairs of numbers and ask your child to determine the rule that relates the numbers. If you name the pairs 1 and 4, 3 and 6, and 10 and 13, your child should determine that the rule is + 3.

5. Name an addition or subtraction fact and ask your child to name other facts in the same fact family. If you name 5 + 4 = 9, your child should say 4 + 5 = 9, 9 – 5 = 4, or 9 – 4 = 5.

6. Practice addition and subtraction by rolling two dice and then adding or subtracting the two numbers shown by the dots. Take turns and have your child check your answers.

7. Set aside about 5 minutes each day for regular practice with Fact Triangles.

8. Name a number and ask your child to tell you how to make that number into a 10. If you say 8, your child should say "add 2 to make 10." If you say 17, your child should say "subtract 7 to make 10."

Building Skills through Games

In Unit 3 your child will practice subtraction facts by playing the following games.

Salute!

Children play in groups of 3. The dealer gives one card to each of two players. Without looking at their cards, the players place them on their foreheads facing out. The dealer finds the sum of the numbers on the cards and says it aloud. Each player uses the sum and the number on the opposing player's forehead to find the number on his or her own card.

The sum is 12.

Subtraction Top-It

Each player draws two cards and subtracts the smaller number from the larger number. The player with the largest difference takes all the cards.

As You Help Your Child with Homework

As your child brings home assignments, you may want to go over the instructions together, clarifying them as necessary. The following answers will guide you through the Unit 3 Home Links.

Home Link 3-1

1. 8 + 2 = 10

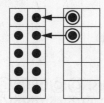

2. 6 + 5 = 11

3. 8 + 5 = 13

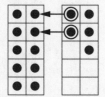

4. 8   **5.** 14   **6.** 9   **7.** 16

Home Link 3-2

1. $6 + 9 = 15$; $9 + 6 = 15$; $15 - 9 = 6$; $15 - 6 = 9$

2. $7 + 8 = 15$; $8 + 7 = 15$; $15 - 8 = 7$; $15 - 7 = 8$

3. $9 + 5 = 14$; $5 + 9 = 14$; $14 - 5 = 9$; $14 - 9 = 5$

4. 10   **5.** 12   **6.** 10   **7.** 11

Home Link 3-4

Round 1: 5   **Round 2:** 6   **Round 3:** 5

Home Link 3-5

1. 3   **2.** 2   **3.** 3   **4.** 2

5. 7   **6.** 9   **7.** 12   **8.** 11

9. counting up

10. counting back
Sample answer: Because 2 is a small number, it's easier to count back 2 and get 11.

11. $6 + 7 = 13$ **12.** $8 + 4 = 12$

Home Link 3-6

1. 7   **2.** 11   **3.** 8   **4.** 0   **5.** 12

6. 9   **7.** 10   **8.** 12   **9.** 18   **10.** 17

Home Link 3-7

1. 15; 17; 14   **2.** -8 **3.** 6; 3; 9; 0

4. $+5$; 18; 5; Answers vary.

Home Link 3-8

1. 7; Sample answer: I know $6 + 6 = 12$, and 13 is 1 more than 12. So I added 1 to one of the 6s. The answer is 7.

2. 9; Sample answer: I know that $8 + 8 = 16$ and 17 is 1 more than 16. So I added 1 to one of the 8s. The answer is 9.

3. 10   **4.** 11

Home Link 3-9

1. 9

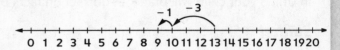

2. 8

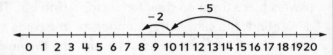

3. $5 + 3 = 8$ **4.** $15 = 6 + 9$

Home Link 3-10

1. 6

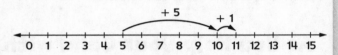

2. 7

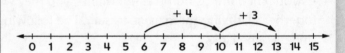

3. 12   **4.** 10   **5.** 14

Home Link 3-11

1. 33¢   **2.** 34¢   **3.** 52¢

4. Sample answers: Ⓠ Ⓠ Ⓝ Ⓟ Ⓟ;
Ⓓ Ⓓ Ⓓ Ⓓ Ⓟ Ⓟ Ⓟ Ⓟ Ⓟ Ⓟ Ⓟ

Addition Strategies

Family Note

In this lesson we learned about the making-10 strategy for adding numbers. We looked at dots arranged in two ten frames and discussed how to figure out the total number without counting each dot one by one. One way is to move dots from one frame in order to fill up the other frame to "make 10." This strategy can make addition easier because many of us are good at adding numbers when 10 is one of them. We will revisit making 10 along with other addition strategies throughout the year.

Please return this Home Link to school tomorrow.

① Show on the double ten frame how to use the making-10 strategy to find the total number of dots. Then write a number model.

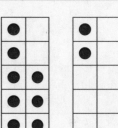

Number model: _____

② Show on the double ten frame how to use the making-10 strategy or a doubles fact to find the total number of dots. Then write a number model.

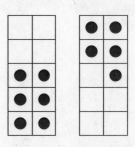

Number model: _____

③ Tell someone at home how to find the total number of dots. Use any strategy except counting by ones. Then write a number model.

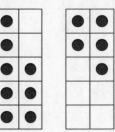

Number model: _____

Practice

Solve.

④ 4 + 4 = _____   ⑤ _____ = 7 + 7

⑥ 4 + 5 = _____   ⑦ _____ = 7 + 9

Domino Facts

Family Note

Today we learned about related addition facts and subtraction facts. For example, $5 + 3 = 8$ has two related subtraction facts: $8 - 5 = 3$ and $8 - 3 = 5$. Each domino shown below can be used to write two addition facts and two related subtraction facts.

Please return this Home Link to school tomorrow.

Write two addition facts and two subtraction facts for each domino.

Example:

$$\frac{\begin{array}{r} 3 \\ + 7 \end{array}}{10} \quad \frac{\begin{array}{r} 7 \\ + 3 \end{array}}{10} \quad \frac{\begin{array}{r} 10 \\ - 7 \end{array}}{3} \quad \frac{\begin{array}{r} 10 \\ - 3 \end{array}}{7}$$

①

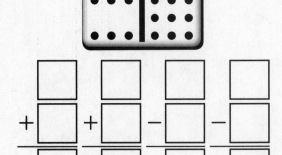

$$\frac{\begin{array}{r} \Box \\ + \Box \end{array}}{\Box} \quad \frac{\begin{array}{r} \Box \\ + \Box \end{array}}{\Box} \quad \frac{\begin{array}{r} \Box \\ - \Box \end{array}}{\Box} \quad \frac{\begin{array}{r} \Box \\ - \Box \end{array}}{\Box}$$

②

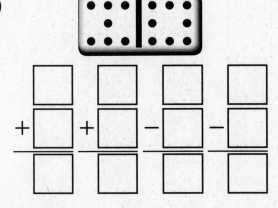

$$\frac{\begin{array}{r} \Box \\ + \Box \end{array}}{\Box} \quad \frac{\begin{array}{r} \Box \\ + \Box \end{array}}{\Box} \quad \frac{\begin{array}{r} \Box \\ - \Box \end{array}}{\Box} \quad \frac{\begin{array}{r} \Box \\ - \Box \end{array}}{\Box}$$

③

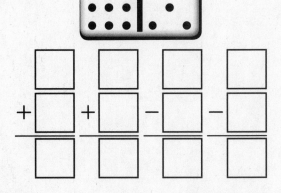

$$\frac{\begin{array}{r} \Box \\ + \Box \end{array}}{\Box} \quad \frac{\begin{array}{r} \Box \\ + \Box \end{array}}{\Box} \quad \frac{\begin{array}{r} \Box \\ - \Box \end{array}}{\Box} \quad \frac{\begin{array}{r} \Box \\ - \Box \end{array}}{\Box}$$

Facts Practice

Fill in the unit box. Solve the problems.

Unit

4. $7 + 3 =$ _____

5. _____ $= 7 + 5$

6. $6 + 4 =$ _____

7. $6 + 5 =$ _____

Fact Triangles

Family Note

Fact Triangles are tools used to help build mental arithmetic skills. You might think of them as the *Everyday Mathematics* version of flash cards. Fact Triangles are more effective than traditional flash cards for helping children memorize facts, however, because of their emphasis on fact families. A fact family is a collection of related addition and subtraction facts that use the same three numbers. The fact family for the numbers 2, 4, and 6 consists of $2 + 4 = 6$, $4 + 2 = 6$, $6 - 4 = 2$, and $6 - 2 = 4$.

To use Fact Triangles to practice addition with your child, cover the number next to the large dot with your thumb.

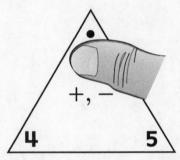

Your child should tell you an addition fact: $4 + 5 = 9$ or $5 + 4 = 9$.

To use Fact Triangles to practice subtraction, cover one of the numbers in the lower corners with your thumb.

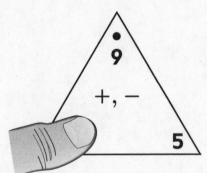 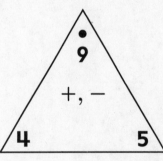

Your child should tell you the corresponding subtraction fact: $9 - 5 = 4$ or $9 - 4 = 5$.

If your child misses a fact, flash the other two fact problems on the card and then return to the fact that was missed.

For example: Sue can't answer $9 - 5$. Flash $4 + 5$, then $9 - 4$, and finally $9 - 5$ a second time.

Make this activity brief and fun. Spend about 5–10 minutes each night over the next few weeks or until your child masters all of the facts. The work that you do at home will help your child develop an instant recall of facts and will complement the work that we are doing at school.

Fact Triangles
(continued)

Cut out the Fact Triangles. Show someone at home how you use them to practice adding and subtracting.

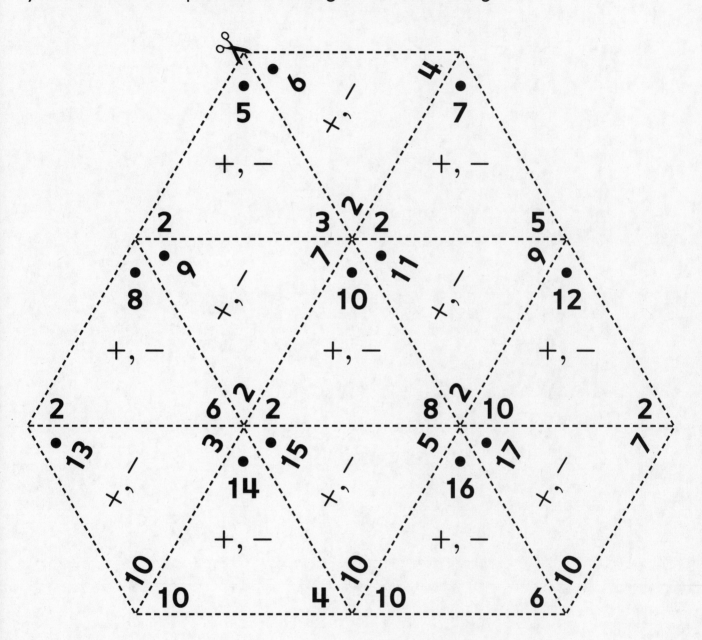

Fact Triangles
(continued)

Cut out the Fact Triangles. Show someone at home how you use them to practice adding and subtracting.

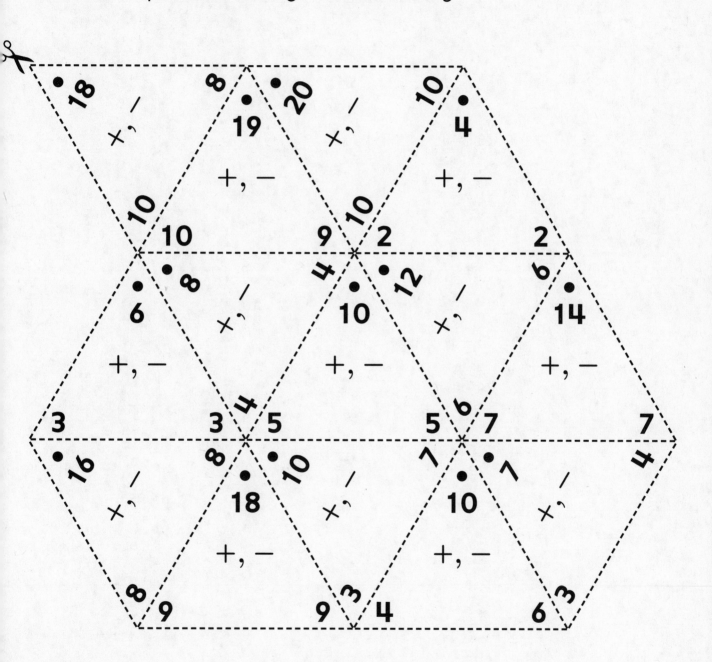

Finding Missing Addends

Family Note

In this lesson children played *Salute!* to practice finding missing addends, which helps them develop fluency with basic facts. Children may solve missing-addend problems by subtracting, counting up, or counting back. *Salute!* is played with three players and a deck of cards. Remove the face cards and jokers before you begin.

$$3 + \boxed{} = 5$$

addend   missing   sum
           addend

Directions

1. One person is selected as the dealer. The dealer gives one card to each of the other two players.

2. Without looking at their cards, players hold them on their foreheads with the number facing out.

3. The dealer looks at both cards and says the sum of the two numbers.

4. Each player looks at the other player's card, keeping in mind the sum said by the dealer. The object is for the players to figure out the number on their own card and say it aloud.

5. After both players have said their numbers, they can look at their own cards.

6. Rotate roles and repeat the game.

7. Play continues until all cards have been used or each player has been the dealer five times.

Please return this Home Link to school tomorrow.

Find the missing addends for the rounds of *Salute!*

MRB 162-163

	Partner 1	**Partner 2**	**Dealer says the sum is ...**
Round 1	5♠ (five of spades)		10
Round 2		2♣ (two of clubs)	8
Round 3		7♠ (seven of spades)	12

Counting Up and Counting Back

Family Note

Today we learned about two subtraction strategies: counting up and counting back. We can use the counting-up strategy when the numbers in a subtraction problem are close together. For example, to solve $11 - 9$ it is easier to start at 9 and count up to 11. This takes 2 counts (9 to 10 and 10 to 11), so the answer is 2. When the number being subtracted is small, the counting-back strategy is easier. For example, to solve $12 - 3$ it is easier to start at 12 and count back 3 (12 to 11, 11 to 10, and 10 to 9). We end on 9, so the answer is 9.

Please return this Home Link to school tomorrow.

Use the counting-up strategy to solve.

(1) $7 - 4 =$ _____   (2) $9 - 7 =$ _____

(3) $11 - 8 =$ _____   (4) $13 - 11 =$ _____

Use the counting-back strategy to solve.

(5) $9 - 2 =$ _____   (6) $12 - 3 =$ _____

(7) $14 - 2 =$ _____   (8) $15 - 4 =$ _____

Write "counting up" or "counting back" on the line.

(9) To solve $13 - 9$, _____ is faster.

(10) To solve $13 - 2$, _____ is faster.

Explain your answer. _____

_____

Practice

Write the turn-around fact for each addition fact.

(11) $7 + 6 = 13$ _____   (12) $4 + 8 = 12$ _____

Using Subtraction Strategies

Family Note

In today's lesson children explored the − 0 (minus 0) and − 1 (minus 1) fact strategies:

- If 0 is subtracted from any number, that number doesn't change.
  Examples: **3** − 0 = **3** and **27** − 0 = **27**

- If 1 is subtracted from any number, the result is the next smaller number.
  Examples: **7** − 1 = **6** (6 is the next smaller number) and **48** − 1 = **47** (47 is the next smaller number)

Children also played *Subtraction Top-It* to practice using subtraction strategies. They will be exposed to various strategies throughout the year to help develop fluency with subtraction facts.

Please return this Home Link to school tomorrow.

Solve.

Unit

① $8 - 1 =$ _____

② _____ $= 11 - 0$

③ $9 - 1 =$ _____

④ $12 -$ _____ $= 12$

⑤ $13 - 1 =$ _____

⑥ _____ $= 10 - 1$

Practice

Solve.

⑦ $5 + 5 =$ _____

⑧ $5 + 7 =$ _____

⑨ _____ $= 9 + 9$

⑩ _____ $= 8 + 9$

"What's My Rule?"

Family Note

Today your child learned about a kind of problem you may not have seen before. We call it a "What's My Rule?" problem. Please ask your child to explain it to you.

Background information: Imagine a machine with an input funnel on top and an output tube at the bottom. The machine is programmed so that if you drop a number into the funnel, the machine does something to the number, and a new number comes out of the tube. *For example:* The machine is set to add 5 to any input number. If you put in 3, 8 comes out. If you put in 7, 12 comes out.

We call this device a function machine.

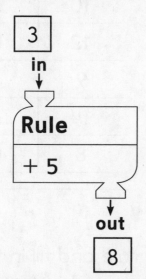

You can show the results of the rule "+ 5" in a table:

in	out
3	8
7	12
15	20

In a "What's My Rule?" problem, some of the information is missing. To solve the problem, you have to find the missing information, which could be the numbers that come out of a function machine, the numbers that are dropped in, or the rule for programming the machine. *For example:*

Rule	
Add 6	
in	**out**
3	
5	
8	

Missing: *out* numbers

Rule	
in	**out**
6	3
10	7
16	13

Missing: *rule*

Rule	
+ 4	
in	**out**
	6
	16
	11

Missing: *in numbers* numbers

Like Frames-and-Arrows problems, "What's My Rule?" problems help children practice facts (and other addition and subtraction problems) in a problem-solving format.

Please return the second page of this Home Link to school tomorrow.

Give the Family Note to someone at home. Show that person how you can complete "What's My Rule?" tables. Show that person how you can find rules.

① Fill in the table.

Rule

+ 9

in	out
1	10
4	13
6	
8	
5	

② Fill the rule.

Rule

in	out
10	2
12	4
9	1
14	6
8	0

③ Fill in the table.

Rule

+ 6

in	out
4	10
	12
	9
	15
	6

Try This

④ Fill the rule and fill in the missing *in* and *out* numbers.

Rule

in	out
8	13
4	9
13	
	10

Finding Missing Addends

Family Note

In this lesson children learned about using doubles facts to help solve subtraction facts. This is a strategy children can use whenever a subtraction fact is related to an addition double or to an addition fact close to a double. To solve $8 - 4 =$ _____, children might think of the related addition double $4 + 4 = 8$ to find the answer 4. To solve $15 - 8 =$ _____, children could think $8 +$ _____ $= 15$. Starting from the double $8 + 8 = 16$, they will find the solution is 1 less than 8, or 7.

Please return this Home Link to school tomorrow.

Look at the missing addend in each Fact Triangle. Tell someone at home how to use doubles to help find it.

Unit

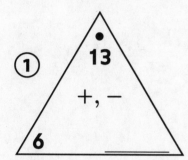

① Explain how you found the missing addend.

_____

_____

_____

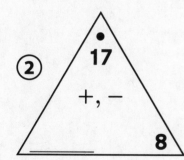

② Explain how you found the missing addend.

_____

_____

_____

Practice

Solve.

③ $8 + 2 =$ _____ ④ $8 + 3 =$ _____

Subtraction Strategy: Going Back Through 10

Family Note

In today's lesson your child learned a subtraction strategy called **going back through 10.** Because 10 is a friendly number that children are comfortable working with, children break the subtraction into two steps, using 10 as a breaking point. To solve $12 - 5 = ?$ your child may say:

- Start from 12, take away 2 to get to 10, and then take away 3 more to get to 7. By taking away the two parts 2 and 3, I get to 7. So the answer is $12 - 5 = 7$.

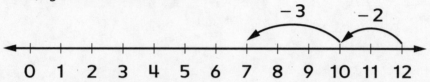

Your child may choose to use a different strategy to subtract, but it's important to expose him or her to various strategies. Help your child solve the problems below by going back through 10.

Please return this Home Link to school tomorrow.

Solve each problem using the going-back-through-10 strategy. Use the number line to show your work. Then explain your thinking.

(1) $13 - 4 =$ _____

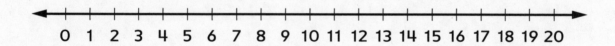

(2) $15 - 7 =$ _____

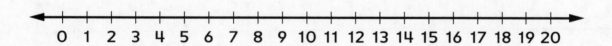

Explain how you solved Problems 1–2 to someone at home.

Practice

Write the turn-around fact.

(3) $3 + 5 = 8$ _____

(4) $15 = 9 + 6$ _____

Going-Through-10 Strategies

Family Note

In today's lesson your child learned a subtraction strategy called going up through 10. This strategy is similar to the going-back-through-10 strategy your child learned in Lesson 3-9. Children use 10 as a friendly number to solve subtraction facts, but they go up instead of back. To solve $13 - 8 = ?$ your child may say the following:

- Start from 8. Go up to 10, which is 2. Then go up to 13, which is 3 more. By adding together the two parts 2 and 3, I get 5. So the answer is $13 - 8 = 5$.

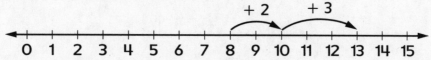

Your child might use a different strategy to subtract, but it's important to expose him or her to a variety of strategies. Parents can help children by guiding them to solve problems using the going-up-through-10 strategy and asking them to explain what they are thinking as they use it.

Please return this Home Link to school tomorrow.

Solve each problem using the going-up-through-10 strategy. Use the number line to show your work. Then explain your thinking to someone at home.

(1) $11 - 5 = $ _____

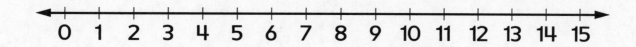

(2) $13 - 6 = $ _____

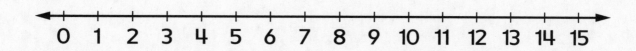

Practice

Solve.

(3) _____ $= 6 + 6$ **(4)** $9 + 1 = $ _____   **(5)** _____ $= 6 + 8$

Counting Coins

Family Note

In this activity your child will count combinations of coins and write each group's value. Children have learned to count coins in order of value: they count quarters first, then dimes, then nickels, and finally pennies. Your child will also draw coins to show different ways to make a certain amount of money. For example, your child might show 28¢ with 1 quarter and 3 pennies or with 2 dimes, 1 nickel, and 3 pennies. You might provide real coins for your child to use for this Home Link.

Please return this Home Link to school tomorrow.

In Problems 1–3, write the total amount.

United States Mint Image

 ①

Total _____¢

②

Total _____¢

③

Total _____¢

④ Show two different ways to make 57¢.
Use Ⓟ, Ⓝ, Ⓓ, and Ⓠ.

Place Value and Measurement

In Unit 4 your child will tell and write times using analog and digital clocks and discuss how to use A.M. and P.M. to specify the time of day.

Children will read, write, and compare numbers from 0 through 999, building on concepts and skills explored in *Everyday Mathematics* for Kindergarten and first grade. They will also review and extend their understanding of place value, which is the system that gives each digit a value according to its position in a number. In the number 52, for example, the 5 represents 5 tens (or 50), and the 2 represents 2 ones (or 2).

Unit 4 also focuses on estimating and measuring lengths using inches, centimeters, and feet. Children will learn that measurements are not exact, and they will use terms such as *close to, a little more than, a little less than, between,* and *about* when describing measurements.

Math Tools

Children will use **base-10 blocks** to help them understand place value. These blocks represent the number 52.

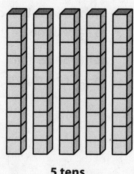

5 tens   **2 ones**

Your child will use rulers marked with standard units to measure length. *Everyday Mathematics* uses both U.S. customary and metric units.

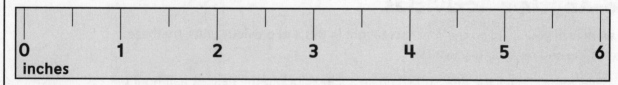

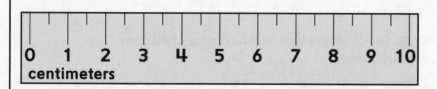

Vocabulary   Important terms in Unit 4:

analog clock A clock that shows time by the position of the hour and minute hands.

digital clock A clock that shows time with numbers of hours and minutes, usually separated by a colon.

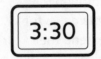

base-10 blocks In *Second Grade Everyday Mathematics,* a set of blocks for representing ones, tens, and hundreds.

cube A base-10 block representing 1 in *Second Grade Everyday Mathematics.*

cube

long A base-10 block representing 10 in *Second Grade Everyday Mathematics.*

flat A base-10 block representing 100 in *Second Grade Everyday Mathematics.*

long

A flat

base-10 shorthand Simple drawings of base-10 blocks used to quickly record work.

Base-10 shorthand

digit Any one of the symbols 0, 1, 2, 3, 4, 5, 6, 7, 8, and 9. Numbers are made up of digits. The number 145, for example, is made up of the digits 1, 4, and 5. In the base-10 number system the value of a digit depends on its place in the number. In the number 145 the digit 1 is worth 100 because it is in the hundreds place.

standard unit A unit of measure that has been defined by a recognized authority, such as a government or a standards organization. Inches, feet, and centimeters are examples of standard units.

foot (ft) A U.S. customary unit of length equal to 12 inches.

inch (in.) A U.S. customary unit of length equal to $\frac{1}{12}$ of a foot.

centimeter (cm) A metric unit of length equal to $\frac{1}{100}$ of a meter.

Do-Anytime Activities

To work with your child on the concepts taught in this and previous units, try these interesting and rewarding activities:

1. Have your child tell the time shown on an analog clock to the nearest half hour or 5 minutes, depending on your child's skill level. By the end of second grade, children are expected to tell time to the nearest 5 minutes.

2. Draw an analog clock without hands. Say or write a time and have your child draw hands in the correct positions on the clock face.

3. Ask your child to tell you the value of a digit in any 3-digit number. In 694, for example, the 6 is worth 600, the 9 is worth 90, and the 4 is worth 4.

4. Name pairs of numbers and ask your child to determine which number is larger.

5. Discuss the different things you could measure with a ruler or a tape measure, such as the length of a book, the height of a table, or the distance from the refrigerator to the sink. Have your child give an estimate of a length or distance before measuring. Record the data and continue periodically to measure things with your child.

Building Skills through Games

In Unit 4 your child will practice mathematical skills by playing the following games.

Evens and Odds

Each player draws a card. If the card shows an even number, the player writes that number as a sum of two equal numbers. (For 6, the child writes $3 + 3 = 6$.) If the card shows an odd number, the player writes that number as the sum of two equal numbers plus or minus 1. (For 7, the child writes $3 + 3 + 1 = 7$ or $4 + 4 - 1 = 7$.)

Addition Top-It

Each player draws two number cards and adds the two numbers. The player with the larger sum takes all four cards.

Number Top-It

Each player uses two or more cards to build a multidigit number. The player with the largest number wins the round.

Target

Players draw number cards to create 1- and 2-digit numbers and use base-10 blocks to represent them. Players add or subtract each new number from their current total until the blocks on one player's mat have a value of exactly 50.

The Exchange Game (with Base-10 Blocks)

Each player rolls a die and collects that number of base-10 cubes from the bank. As players accumulate cubes, they exchange 10 cubes for 1 long. As they accumulate longs, they exchange 10 longs for 1 flat.

As You Help Your Child with Homework

When your child brings home assignments, you may want to go over the instructions together, clarifying them as necessary. The following answers will guide you through the Unit 4 Home Links.

Home Link 4-1

1. 4:00   **2.** 8:30   **3.** 11:30

4.   **5.**   **6.**

7. 11   **8.** 4   **9.** 6   **10.** 16

Home Link 4-2

1. Answers vary.   **2.** 6:30

3. 9:40   **4.** 1:25   **5.** 2:15

6.   **7.**

8.   **9.**

10. 12   **11.** 8   **12.** 7   **13.** 13

Home Link 4-3

1. Answers vary.

2. 9   **3.** 7   **4.** 2   **5.** 9

Home Link 4-4

1. a. 374   **b.** 507   **2.** 740

3. 936   **4.** 8; 0; 6   **5.** 2; 3; 5

Home Link 4-5

Answers vary for the rounds of *Number Top-It*.

1. 14   **2.** 16   **3.** 8   **4.** 7

Home Link 4-6

1. 145   **2.** 328

Home Link 4-7

1. 32; 17; 49

Sample answer:

2. 26; 34; 60

Sample answer:

3. 32

Sample answer:

29

Home Link 4-8

Answers vary.

Home Link 4-9

1. Answers vary.

2. 14   **3.** 6   **4.** 13   **5.** 13

Home Link 4-10

1. Answers vary.

2. 8   **3.** 5   **4.** 9   **5.** 15

Home Link 4-11

12 cm; 6 cm; 8 cm; 26 cm

Telling Time to the Half Hour

Family Note

Today we reviewed telling and writing times to the hour and half hour on an analog clock. We discussed the movement of the hour and minute hands and units of time such as hour and minute.

Please return this Home Link to school tomorrow.

Write the time shown on each clock.

MRB
107

①

②

③

_____ : _____   _____ : _____   _____ : _____

Draw the hour hand and the minute hand to show the time.

④

9:30

⑤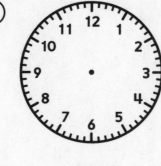

3:30

⑥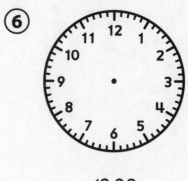

12:30

Practice

⑦ 6 + 5 = _____

⑧ _____ + 3 = 7

⑨ 7 + _____ = 13

⑩ _____ = 8 + 8

Times of Day

Family Note

Your child is learning how to tell time by writing times displayed on an analog clock (a clock with an hour hand and a minute hand) and by setting the hands on an analog clock to show specific times. Your child should have brought home a clock to use while completing the exercises on this page. Ask your child to use the clock to show you other times that don't appear here.

Please return this Home Link to school tomorrow.

① Use your clock to show someone at home the time you do the following activities. Write the time under each activity.

MRB
107

Eat dinner   Go to bed   Get up   Eat lunch

_____:_____   _____:_____   _____:_____   _____:_____

Write the time.

②

_____:_____

③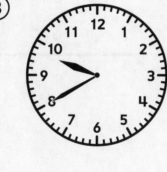

_____:_____

④

_____:_____

⑤

_____:_____

seventy-five   75

Times of Day (continued)

Draw the hands to match the time.

⑥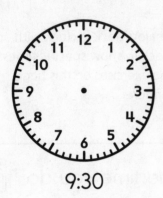

9:30

⑦

4:05

⑧

12:45

⑨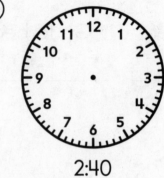

2:40

Practice

Solve the facts.

⑩ 7 + 5 = _____

⑪ _____ + 6 = 14

⑫ 3 + _____ = 10

⑬ 4 + 9 = _____

A.M. and P.M.

Family Note

Today we discussed the meanings of A.M. and P.M. Your child learned that A.M. describes times from 12:00 midnight to 12:00 noon and that P.M. describes times from 12:00 noon to 12:00 midnight. We identified events that occur throughout the day and labeled them on a 24-hour timeline.

Talk with your child about events that take place during your family's day, such as eating dinner, doing homework, reading, getting ready for bed, sleeping, and waking up.

Please return this Home Link to school tomorrow.

① Draw pictures of things that happen at home.

Write the time for each using A.M. and P.M.

Time: _____	Time: _____
Time: _____	Time: _____

Practice

② $6 + 3 =$ _____

③ $5 +$ _____ $= 12$

④ _____ $+ 8 = 10$

⑤ _____ $= 5 + 4$

Place Value

Family Note

All numbers are made up of digits. A digit's value depends on its place in the number. In the number 704, the digit 7 means 7 hundreds, the digit 0 means 0 tens, and the digit 4 means 4 ones. This idea is called *place value*. Your child has been using base-10 blocks to work with place value. Base-10 blocks are shown in Problem 1. A "cube" (with each side 1 unit long) represents 1. A "long" (a rod that is 10 units long) represents 10. And a "flat" (a square with each side 10 units long) represents 100.

Please return this Home Link to school tomorrow.

① What number do the base-10 blocks show?

MRB
71–73

a.

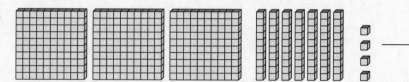

_____

b.

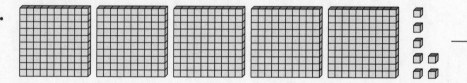

_____

② Write a number with
7 in the hundreds place,
0 in the ones place, and
4 in the tens place.

_____

③ Write a number with
3 in the tens place,
6 in the ones place, and
9 in the hundreds place.

_____

④ In 806, there are

_____ hundreds,

_____ tens, and

_____ ones.

⑤ In 235, there are

_____ hundreds,

_____ tens, and

_____ ones.

Comparing Numbers

Family Note

Today we practiced comparing numbers by playing a game called *Number Top-It*. (*See directions below.*) You can make cards by writing the numbers 0–9 on index cards (make four cards for each number), or you can use a deck of playing cards. If you use playing cards, you will first need to change the four queens to 0s, change the four aces to 1s, and remove the jacks, kings, and jokers.

Please return this Home Link to school tomorrow.

Play *Number Top-It* with someone at home:

MRB
170–172

① Shuffle the cards. Place the deck number-side down.

② Take turns drawing cards until each player has three cards.

③ Each player uses their three cards to make a 3-digit number and reads the number aloud.

④ Compare the two numbers. The player with the larger number for the round scores 1 point, and the player with the smaller number scores 2 points.

⑤ Play five rounds per game. When you've used all the cards in the deck, shuffle them to make a new deck. The player with the fewest points at the end of five rounds wins the game.

Practice

Solve the facts.

① $5 + 9 =$ _____

② _____ $= 9 + 7$

③ _____ $+ 4 = 12$

④ $2 +$ _____ $= 9$

Using Base-10 Blocks

Family Note

Today we explored how to write numbers shown by base-10 blocks. In this lesson, we decided that making a trade can sometimes help us find the number the blocks represent. Making trades with base-10 blocks will be revisited throughout the year in games and addition and subtraction situations.

Please return this Home Link to school tomorrow.

(1) Shane had these base-10 blocks:

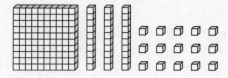

He made a trade. Then he showed the same number in base-10 shorthand:

☐ | | | | | ▪ ▪ ▪ ▪ ▪   The number shown here is

_____.

(2) Suppose you have these base-10 blocks:

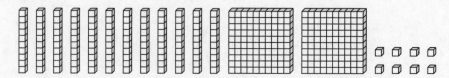

Make a trade. Then draw the base-10 blocks and write the number shown. See the example in Problem 1. The number shown here is _____.

Making Exchanges

Family Note

Today your child used base-10 blocks to represent, add, and subtract 2-digit numbers. When adding, children often exchange 10 ones for 1 ten to represent the final number using the fewest possible blocks. When subtracting, children often need to exchange 1 ten for 10 ones to have enough ones to take away. Ask your child to explain how they represent numbers for the problems below.

Please return this Home Link to school tomorrow.

Write the numbers shown by the blocks.

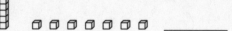
long   cube

① _____   _____

What is the total value? _____

Use base-10 shorthand to show your answer:

② _____   _____

What is the total value? _____

Use base-10 shorthand to show your answer:

③ _____

Use base-10 shorthand to show how you can take away
3 cubes. *Hint:* Exchange 1 long for 10 cubes.

What is the value of the blocks that are left? _____

Talk to someone at home about making exchanges
between base-10 longs and cubes.

Measuring with a Foot-Long Foot

Family Note

Today we talked about the importance of measuring with standard units so that we all get the same results. You and your child can use the foot-long (12-inch) foot to measure objects or distances around your home. Objects or distances will usually be longer or shorter than a whole number of feet, so encourage your child to use language such as "about _____ feet," "a little less/more than _____ feet," or "about halfway between _____ and _____ feet."

Please return this Home Link to school tomorrow.

Follow these steps:

① Cut out the foot-long foot.

② Measure three objects or distances to the nearest foot. Write your measurements in the chart.

③ Have someone else measure the same things. Write their measurements in the chart.

④ Agree on a measurement that is close.

Object or Distance	My Measurements	Another Person's Measurements
Example: table	between 6 feet and 7 feet	between 6 feet and 7 feet

Measuring in Inches

Family Note

In today's lesson your child learned to measure objects in inches with a 12-inch (foot-long) ruler. We also discussed the important concept that an inch ruler is composed of a series of inch-long spaces. We measured short objects first with inch-long blocks and then with 12-inch rulers to show that the measurements are the same.

Please return this Home Link to school tomorrow.

① Cut out the 6-inch ruler below. Measure four short objects or distances to the nearest inch. Record your measures below.

MRB
101

Object or Distance	Length to the Nearest Inch
	About _____ inches
	About _____ inches
	About _____ inches
	About _____ inches

Practice

Solve the facts.

② 6 + 8 = _____

③ 9 + _____ = 15

④ 7 + 6 = _____

⑤ _____ = 8 + 5

Measuring in Centimeters

Family Note

Today your child learned about the metric unit of length called the centimeter. The inch, introduced in the previous lesson, is a length unit in the U.S. customary system of measurement. With the exception of the United States, most countries use the metric system in everyday life. People in the United States and the rest of the world use the metric system for scientific purposes. It is important for your child to become proficient in both measurement systems.

Please return this Home Link to school tomorrow.

① Cut out the 10-centimeter ruler below. Measure three short objects or distances to the nearest centimeter. Record your measurements in the table.

Object or Distance	Length
	About _____ centimeters
	About _____ centimeters
	About _____ centimeters

Practice

Solve the facts.

② 6 + 2 = _____

③ 7 + _____ = 12

④ _____ + 3 = 12

⑤ _____ = 7 + 8

Unit

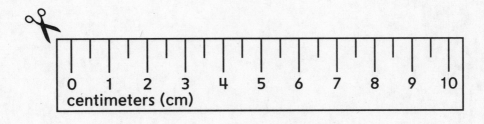

Measuring to the Nearest Centimeter

Family Note

In today's lesson your child measured the length of a long path to the nearest inch and the nearest centimeter (cm). Ask your child to explain how to measure each section of the path on this page. Encourage your child to measure objects at home. If you don't have a ruler at home, have your child cut out and use the 10-centimeter ruler at the bottom of the page.

Please return this Home Link to school tomorrow.

A ladybug walked around the garden. Measure each part of its path to the nearest centimeter. Record the measurements in the table.

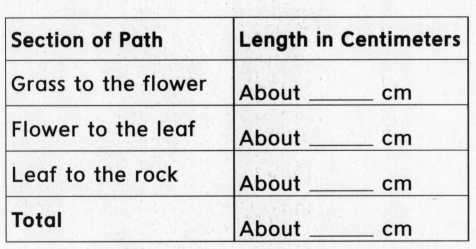

Section of Path	Length in Centimeters
Grass to the flower	About _____ cm
Flower to the leaf	About _____ cm
Leaf to the rock	About _____ cm
Total	About _____ cm

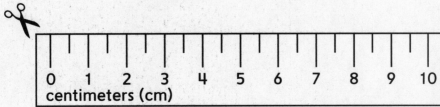

0 1 2 3 4 5 6 7 8 9 10
centimeters (cm)

Unit 5: Family Letter

Addition and Subtraction

In Unit 5 your child will review and extend money concepts. The class will find the total value of combinations of coins, find different coin combinations that have the same total value, and make change.

Your child will also develop mental arithmetic skills, or computations that children do in their heads. As they develop mental arithmetic skills, children may draw pictures or use various tools—such as counters, money, number lines, and number grids—to help them solve problems. In this unit children use a new tool, the **open number line,** to record their mental strategies for adding and subtracting 2-digit numbers. Home Link 5-7 will include more information about open number lines.

> I know that 34 is three 10s and four 1s. I start at 42 and add three 10s: 42 plus 10 is 52, plus 10 more is 62, plus 10 more is 72. Now I need to add 4 more: 72 plus 4 is 76.

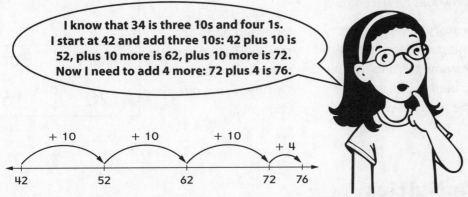

A second grader uses an open number line to solve 42 + 34.

At the end of this unit, children will solve addition and subtraction number stories. Two basic types of addition situations are change-to-more and putting together. Children will use **change diagrams** and **parts-and-total diagrams** to help organize information in addition stories that either "change to more" or "put together." They will also use change diagrams to organize information in stories about temperature changes, which may be either change-to-more (addition) or change-to-less (subtraction) stories. See the Vocabulary section in this Family Letter to see examples and learn more about these diagrams.

Please keep this Family Letter for reference as your child works through Unit 5.

Vocabulary   Important terms in Unit 5:

open number line A blank number line on which children can mark points or numbers that are useful for solving problems. Children can use open number lines to record the steps of mental computation strategies. *For example:* I want to solve 56 + 28. I can start at 56 and jump up 4 ones to get to an easy number, 60. I still have 24 more to go. Next I can jump up two 10s, to 70 and then to 80. Now I just have four more 1s to go, so I hop 4 to 84. So 56 + 28 = 84.

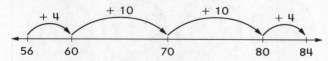

change-to-more number story A number story in which a starting quantity is increased so that the ending quantity is more than the starting quantity. *For example:* Nick has 20 comic books. He buys 6 more. How many does he have now?

change-to-less number story A number story in which a starting quantity is decreased so that the ending quantity is less than the starting quantity. *For example:* Abby has 12 berries. She eats 5 of them. How many does she have now?

change diagram A diagram that organizes information from a change-to-more or change-to-less number story. The following change diagram organizes the information from Nick's comic book story.

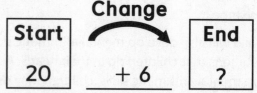

parts-and-total number story A number story in which two or more quantities (parts) are combined to form a total quantity. *For example:* Carl filled 20 gift bags. Sam filled 16 gift bags. How many gift bags did Carl and Sam fill in all?

parts-and-total diagram A diagram that organizes information from a parts-and-total number story. The following parts-and-total diagram organizes the information from Carl and Sam's gift bag story.

Total	
?	
Part	**Part**
20	16

Do-Anytime Activities

To work with your child on the concepts taught in this unit and previous units, try these interesting and rewarding activities:

1. Challenge your child to solve an addition or a subtraction fact faster than you can solve it on a calculator.

2. At the grocery store, show your child an item that costs less than $1. Ask your child what coins or bills he or she would use to pay for the item and how much change the cashier would give back.

3. Pose addition or subtraction problems for your child to solve mentally. Encourage your child to draw an open number line to show his or her problem-solving steps.

4. Look at weather reports in the newspaper, on television, or online. Have your child figure out the difference between the high and low temperatures for each day.

5. Look at temperatures at different points during the day. Ask your child to determine whether the temperature has changed to more or changed to less.

Building Skills through Games

In Unit 5 your child will play the following games to practice solving facts, exchanging coins, and adding and subtracting mentally and with tools.

Beat the Calculator

One player is the Caller, who names two 1-digit numbers. Another player is the Brain, who adds the two numbers mentally. A third player is the Calculator, who adds the numbers with a calculator. The Brain tries to find the sum faster than the Calculator.

Spinning for Money

Players take turns spinning a spinner and taking the indicated coins from the bank. Whenever they can, players exchange their coins for coins in larger denominations (for example, 5 pennies for 1 nickel). The first player to exchange coins for a $1 bill wins.

Target

Players draw number cards to create 1- and 2-digit numbers and use base-10 blocks to represent them. Players add or subtract each new number from their current total until the blocks on one player's mat have a value of exactly 50.

Addition/Subtraction Spin

Players spin a spinner to determine a 3-digit number. Then they roll a die to see if they should add 10 or 100 to the 3-digit number or subtract 10 or 100 from it. Players do the computation mentally.

As You Help Your Child with Homework

As your child brings home assignments, you may want to go over the instructions together, clarifying them as necessary. The answers listed below will guide you through the Unit 5 Home Links.

Home Link 5-1

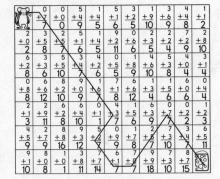

Home Link 5-2

1–3. Answers vary.

4. 9   **5.** 7   **6.** 13   **7.** 16

Home Link 5-3

1–4. Answers vary.

5. 3   **6.** 7   **7.** 7   **8.** 5

Home Link 5-4

5¢; 35¢; 16¢; 5¢; 2¢; 52¢; Answers vary.

1. 3   **2.** 7   **3.** 3   **4.** 6

Home Link 5-5

1. 8:30   **2.** 4:15   **3.** 1:40

4. 7:10   **5.** 11   **6.** 4

Home Link 5-6

1. 72   **2.** 48   **3.** 126

4. 381   **5.** 886   **6.** 525

7. 34   **8.** 205   **9.** 9

10. 7   **11.** 6   **12.** 3

Home Link 5-7

1. 58; Sample number line:

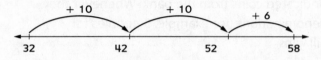

2. 33; Sample number line:

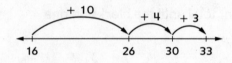

Home Link 5-8

1.

Start	Change	End
11	+ 7	?

11 + 7 = ?; 18 grapes

2.

Start	Change	End
30	+ 8	?

30 + 8 = ?; 38 cards

3.

Start	Change	End
42	+ 10	?

42 + 10 = ?; 52 pounds

Home Link 5-9

1.

Total	
?	
Part	Part
17	30

17 + 30 = ?; 47 pounds

2.

Total	
?	
Part	Part
45	30

45 + 30 = ?; 75 pounds

3.

Total	
?	
Part	Part
17	15

17 + 15 = ?; 32 pounds

Home Link 5-10

1.

Start	Change	End
30	+ ?	42

30 + ? = 42; 12°F

2.

Start	Change	End
65	– ?	50

65 – ? = 50; 15°F

3. Sample answer: I counted up from 50 to 65 and got 15.

Home Link 5-11

1. Strategies vary; $50

2. Strategies vary; $50

Solving Addition Facts

Family Note

Today we continued working with addition facts. Children can develop number-fact reflexes the same way that they develop any other habit—by practicing them over and over. In *Everyday Mathematics* knowing facts automatically is called fact power. We discussed ways to develop fact power, such as practicing with Fact Triangles and playing fact games.

When your child has solved the addition facts below and is ready to draw the mouse's path through the maze, explain that the mouse can move up, down, left, right, or diagonally to find the cheese.

Please return this Home Link to school tomorrow.

Solve the facts. Then draw a path for the mouse to find the cheese. The mouse can go through only those boxes with a sum of 7.

MRB
40–45

	0 +7	0 +0	5 +4	1 +4	5 +1	3 +2	1 +9	3 +6	4 +4	1 +1
	2 +0	3 +5	2 +5	5 +1	1 +4	9 +2	0 +6	2 +3	2 +2	7 +2	2 +8
	6 +2	3 +3	5 +5	3 +4	4 +2	0 +5	1 +8	4 +6	5 +3	4 +0	3 +1
	0 +8	6 +6	8 +2	9 +0	1 +6	7 +1	6 +6	1 +3	1 +5	6 +0	0 +4
	2 +1	2 +9	6 +2	6 +4	0 +1	4 +3	1 +5	6 +3	0 +7	0 +2	1 +2
	4 +5	2 +7	8 +8	9 +3	5 +2	0 +9	1 +7	2 +5	7 +3	3 +4	6 +5
	9 +1	8 +0	1 +0	3 +8	7 +7	6 +1	7 +0	9 +9	7 +3	8 +7	

Paying for Items

Family Note

In class today we reviewed coin equivalencies and found different coin combinations to represent the same amount of money. For example, 12¢ can be shown with 12 pennies, with 2 nickels and 2 pennies, with 1 nickel and 7 pennies, or with 1 dime and 2 pennies. In this activity your child looks through advertisements, selects items that cost less than $2, and shows how to pay for those items by drawing coins and bills. If you do not have access to advertisements, make up some items and prices.

Please return this Home Link to school tomorrow.

Look at newspaper or magazine ads. Find three items that cost less than $2. Write the name and the price of each item. Show someone at home how you could pay for these items with coins and a $1 bill. Write Ⓟ, Ⓝ, Ⓓ, Ⓠ, and $1 .

MRB
110–111

① I would buy _____. It costs _____.

This is how I would pay:

② I would buy _____. It costs _____.

This is how I would pay:

③ I would buy _____. It costs _____.

This is how I would pay:

Practice

Fill in the unit box. Solve.

Unit

④ $17 - 8 =$ _____

⑤ $6 +$ _____ $= 13$

⑥ _____ $- 4 = 9$

⑦ $9 + 7 =$ _____

Change at a Garage Sale

Family Note

Today your child practiced making change by counting up. *For example:* Suppose an apple costs 17¢ and is paid for with a quarter (or 25¢). One way to make change by counting up is to put down three pennies as you say "18, 19, 20" and then put down a nickel and say "25 cents," making 8¢ in change.

In today's Home Link activity your child "sells" small items from around your home at a mock garage sale. Using real coins will make this activity easier. If you feel your child is ready, you can increase the cost of some items and use combinations of coins to pay for them.

Please return the second page of this Home Link to school tomorrow.

Pretend you are having a garage sale. Do the following:

- Find small items in your home to "sell."

- Give each item a different price. Every price should be less than 25¢.

- Pretend that customers pay for each item with a quarter.

- Show someone at home how you would make change by counting up. Use Ⓠ, Ⓓ, Ⓝ, and Ⓟ to draw the change.

Example:

The customer buys ___*a pen*___ for _*21¢*_.

The change is ___Ⓟ Ⓟ Ⓟ Ⓟ___.

Change at a
Garage Sale (continued)

(1) The customer buys _____ for _____.

The change is _____.

(2) The customer buys _____ for _____.

The change is _____.

(3) The customer buys _____ for _____.

The change is _____.

(4) The customer buys _____ for _____.

The change is _____.

Practice

Fill in the unit box. Solve.

(5) $11 - \underline{\hspace{1cm}} = 8$

(6) $8 + \underline{\hspace{1cm}} = 15$

(7) $\underline{\hspace{1cm}} + 7 = 14$

(8) $13 - 8 = \underline{\hspace{1cm}}$

Unit

Counting Up to Make Change

Family Note

Help your child identify the change he or she would receive by counting up from the price of the item to the amount of money used to pay for it. Use real coins and bills to act out the problems with your child. You will need a $1 bill and at least 3 quarters, 5 dimes, 5 nickels, and 5 pennies.

Please return this Home Link to school tomorrow.

Complete the table.

I Buy	It Costs	I Pay With	My Change
A box of raisins	70¢	Q Q Q	_____ ¢
A box of crayons	65¢	$1	_____ ¢
A pen	59¢	Q Q Q	_____ ¢
An apple	45¢	D D D D D	_____ ¢
A notebook	73¢	Q Q D N	_____ ¢
A ruler	48¢	$1	_____ ¢
_____	_____	_____	_____ ¢

Practice

Solve.

① 12 − _____ = 9

② 9 + _____ = 16

③ _____ + 8 = 11

④ 14 − 8 = _____

Unit

Clock Faces and Digital Notation

Family Note

Today your child played *Clock Concentration,* a game that involves matching clock faces to times in digital notation (such as 6:00 or 12:30). By the end of Grade 2, your child is expected to tell time to the nearest 5 minutes. By the end of Grade 3, your child will be expected to tell time to the nearest minute.

Please return this Home Link to school tomorrow.

Draw a line matching each clock face to a time.

①    4:15

②    1:40

③    7:10

④    8:30

Practice

⑤ _____ = 5 + 6   ⑥ 12 − _____ = 8

Unit

Adding and Subtracting 10 and 100

Family Note

Today we learned rules for adding and subtracting 10:

- To add 10, increase the tens digit of a number by 1: $24 + 10 = 34$ $772 + 10 = 782$
- To subtract 10, decrease the tens digit of a number by 1: $98 - 10 = 88$ $615 - 10 = 605$

When the number has a 9 in the tens place (for addition) or 0 in the tens place (for subtraction), the hundreds digit also changes:

- To add 10, increase the hundreds digit by 1 and change the tens digit to 0: $396 + 10 = 406$
- To subtract 10, decrease the hundreds digit by 1 and change the tens digit to 9: $703 - 10 = 693$

We also learned rules for adding and subtracting 100:

- To add 100, increase the hundreds digit of a number by 1: $643 + 100 = 743$
- To subtract 100, decrease the hundreds digit of a number by 1: $451 - 100 = 351$

These rules help children mentally add or subtract 10 or 100.

Please return this Home Link to school tomorrow.

Solve mentally. Tell someone at home about the rules you used.

① $62 + 10 =$ _____

② $58 - 10 =$ _____

③ $116 + 10 =$ _____

④ _____ $= 391 - 10$

⑤ _____ $= 786 + 100$

⑥ $625 - 100 =$ _____

⑦ Clare did 24 sit-ups. She rested and then did 10 more. How many sit-ups did she do in all? _____ sit-ups

⑧ Freddie had 215 marbles. He gave 10 to a friend. How many does he have left? _____ marbles

Practice

⑨ $3 +$ _____ $= 12$

⑩ $16 - 9 =$ _____

⑪ $14 =$ _____ $+ 8$

⑫ $11 -$ _____ $= 8$

Using Open Number Lines

Family Note

Today your child learned about open number lines. Children can use open number lines to quickly record their thinking when they use mental strategies to add or subtract.

Here is an example: To solve 29 + 36, think of 36 as three 10s and six 1s. Add the 10s first. Think, "29 plus 10 is 39, plus 10 more is 49, plus 10 more is 59."

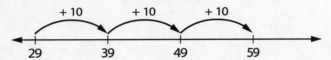

Then add the 1s. Think, "If I add 1 more, I get to 60. Then I still have 5 to go; 60 plus 5 is 65."

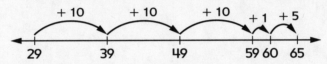

Open number lines are quick and easy to draw because they show only the numbers needed to solve a particular problem. For example, the open number line above only shows 29, 39, 49, 59, 60, and 65 because these are the stopping points used in the mental addition strategy described above.

Please return this Home Link to school tomorrow.

Solve. You may use the open number lines to help.

MRB
78

(1) There are 32 beads on one necklace and 26 beads on another. How many beads in all? _____ beads

(2) You have 16 apples in your basket. You pick 17 more. How many do you have now? _____ apples

Change Number Stories

Family Note

Your child has learned how to represent a problem by using a change diagram, which is shown in the example below. Using diagrams like this can help children organize the information in a problem. When the information is organized, it is easier to decide which operation (+, −, ×, ÷) to use to solve the problem. Change diagrams are used to represent problems in which a starting quantity is increased or decreased. For the number stories on this Home Link, the starting quantity is always increased.

Please return the second page of this Home Link to school tomorrow.

Do the following for each number story on the next page:

- Write the numbers you know in the change diagram.

- Write ? for the number you need to find.

- Write a number model. Use ? for the number you need to find.

- Answer the question.

Example: Twenty-five children are riding on a bus. At the next stop, 5 more children get on. How many children are on the bus now?

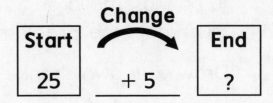

The number of children on the bus has increased by 5.

Possible number model: 25 + 5 = ?

Answer: There are now 30 children on the bus.

Change Number Stories (continued)

① Becky ate 11 grapes after lunch. She ate 7 more grapes after dinner. How many grapes did she eat in all?

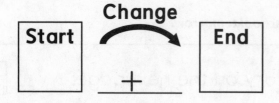

Number model:

_____

_____ grapes

② Bob has 30 baseball cards. He buys 8 more. How many baseball cards does Bob have now?

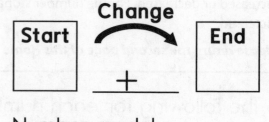

Number model:

_____

_____ cards

③ A large fish weighs 42 pounds. A small fish weighs 10 pounds. The large fish swallows the small fish. How much does the large fish weigh now?

Draw your own change diagram.

Number model: _____

_____ pounds

114   one hundred fourteen

Parts-and-Total Number Stories

Family Note

Your child has learned how to represent and solve problems by using parts-and-total diagrams. Parts-and-total diagrams are used to organize the information in problems in which two or more quantities (parts) are combined to form a total quantity.

Please return the second page of this Home Link to school tomorrow.

Large suitcase 45 pounds

Small suitcase 30 pounds

Backpack 17 pounds

Package 15 pounds

Use the weights shown in the pictures above to do the following for each number story on the next page:

- Write the numbers you know in a parts-and-total diagram.

- Write ? for the number you need to find.

- Write a number model. Use ? for the number you need to find.

- Answer the question.

Example: You carry the small suitcase and the package. How many pounds do you carry in all?

The parts are known. The total is to be found.

Possible number model: $30 + 15 = ?$

Answer: __45__ pounds

Total	
?	
Part	**Part**
30	15

Parts-and-Total Number Stories

(continued)

① You wear the backpack and carry the small suitcase. How many pounds do you carry in all?

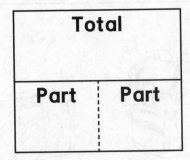

Number model:

_____

Answer: _____ pounds

② You carry the large suitcase and the small suitcase. How many pounds do you carry in all?

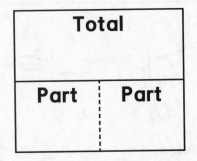

Number model:

_____

Answer: _____ pounds

③ You wear the backpack and carry the package. How many pounds do you carry in all?

Draw your own parts-and-total diagram:

Number model: _____

Answer: _____ pounds

116   one hundred sixteen

Temperature

Family Note

In today's lesson your child solved problems involving temperatures. Thermometers provide a real-world context for solving problems involving change, such as an increase (a change to more) or a decrease (a change to less) in temperature. Change diagrams help children organize information and find the change in a change problem.

On the thermometers on these Home Link pages, the longest degree marks are spaced at 10-degree intervals, the shortest marks are spaced at 1-degree intervals, and the mid-length marks are spaced at 2-degree intervals. Point to these mid-length degree marks while your child counts by 2s: 30, 32, 34, 36, 38, 40, 42, 44 degrees.

Please return the second page of this Home Link to school tomorrow.

For Problems 1–2 on the next page, follow these steps:

- Decide whether the change in temperature is a change to more or a change to less.

- Fill in the diagram with numbers from the problem. Use ? for the number you want to find.

- Write a number model. Use ? for the number you want to find.

- Find the change in temperature.

Example:

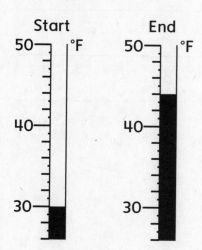

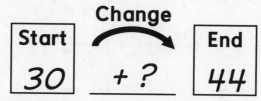
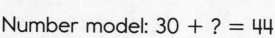

Number model: 30 + ? = 44

Answer: 14°F

Unit

°F

Temperature (continued)

①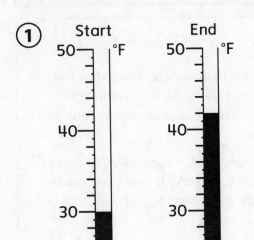

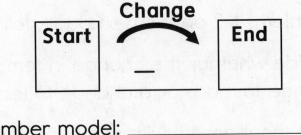

Number model: _____

Answer: _____°F

②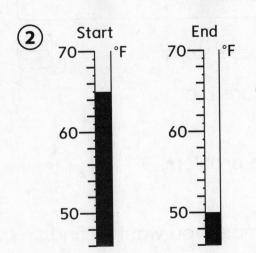

Number model: _____

Answer: _____°F

③ Explain how you found the answer to Problem 2.

_____

_____

_____

Addition Strategies

Family Note

In this lesson we added multidigit numbers. Your child solved an addition number story using two different strategies. Being able to solve problems more than one way and with different tools can help children confirm their answers and choose methods that work well in certain situations. Adding multidigit numbers will be revisited throughout the year.

Please return this Home Link to school tomorrow.

Uma bought a telephone for $36 and blank CDs for $14. What was her total cost?

(1) Show how to solve this problem using base-10 blocks.

Answer: _____

(2) Show how to solve this problem using an open number line.

⟵————————————————————⟶

Answer: _____

Whole Number Operations and Number Stories

In Unit 6 children collect data about the number of pockets on their clothing and display the data in a picture graph (shown below at left) and a bar graph (right).

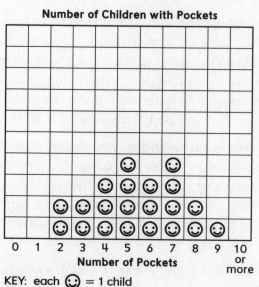

Number of Children with Pockets

Number of Pockets

KEY: each 😊 = 1 child

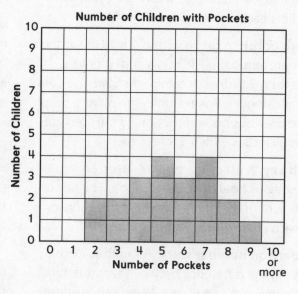

Number of Children with Pockets

Number of Children

Number of Pockets

Children also continue solving number stories and learn to use a new diagram, the **comparison diagram,** to organize information from number stories that involve comparing two different quantities. The comparison diagram at the right shows the information from this comparison number story:

> Barbara has 35 markers. Edward has 20 markers. How many more markers does Barbara have than Edward?

Children also revisit the diagrams introduced in Unit 5, using them to organize their thinking and plan their strategies for solving one- and two-step number stories. Organizing information from a given number story in one of these diagrams can help children decide, for example, whether they should add or subtract to solve the number story.

Quantity
35

Quantity	
20	?
	Difference

A comparison diagram

Throughout the first part of Unit 6, children practice writing number models for number stories using ? to represent the number they need to find. For example, a number model for the number story about Barbara's and Edward's markers might be $20 + ? = 35$.

In the final part of this unit, children invent and use their own strategies to add 2- and 3-digit numbers and are introduced to a formal addition strategy called **partial-sums addition.** Home Links 6-6, 6-7, and 6-8 provide more information about the various addition strategies your child will encounter.

Please keep this Family Letter for reference as your child works through Unit 6.

Vocabulary Important terms in Unit 6:

bar graph A graph with horizontal or vertical bars that represent data. The heights (or lengths) of the bars show the counts for each category. For example, the bar graph on the previous page shows that 4 children are wearing clothes with 5 pockets each.

picture graph A graph with pictures or symbols that represent data. The number of pictures above (or next to) each category shows the count for that category. For example, the picture graph on the previous page shows that 3 children are wearing clothes with 6 pockets each.

graph key A list of the symbols used on a graph that explains how to read the graph. The key on the picture graph on the previous page shows that each smiley-face symbol stands for 1 child.

comparison number story A number story involving the difference between two quantities. *For example:* Ross squeezed 12 lemons. Anthony squeezed 5 lemons. How many more lemons did Ross squeeze than Anthony?

comparison diagram A diagram used to organize information from a comparison number story. For example, the diagram at the right organizes the information from Ross and Anthony's lemon story.

Quantity
12

Quantity	
5	5
	Difference

two-step number story A number story that most children solve using two arithmetic operations. *For example:* Kyla had 6 leaves. She found 8 more in the woods. Then she gave 3 to her sister. How many leaves does Kyla have now?

ballpark estimate A rough estimate that is reasonable or "in the ballpark." Children can use ballpark estimates to check the reasonableness of answers they find using other computation methods. A ballpark estimate for the problem $23 + 81$ might be 100 because $20 + 80 = 100$.

partial-sums addition An addition strategy in which separate sums are computed for each place-value column that are then added to get a final sum. More information on partial-sums addition will be provided in the Family Note for Home Link 6-8.

expanded form A way of writing a number as the sum of the values of its digits. For example, the expanded form of 356 is $300 + 50 + 6$.

Do-Anytime Activities

To work with your child on the concepts taught in this unit and previous units, try these interesting and rewarding activities:

1. Encourage your child to show you his or her favorite addition strategy.

2. Ask your child to make a ballpark estimate for the sum of two 2- or 3-digit numbers.

3. Pose one- and two-step number stories for your child to solve. Ask your child to explain his or her solution strategy to you.

4. Have your child compare two objects' lengths. Ask which object is longer and prompt your child to use a ruler or a tape measure to find the difference between the lengths.

Building Skills through Games

In Unit 6 your child will practice mathematical skills by playing the following games.

The Exchange Game

Each player rolls a die and collects that number of base-10 cubes from the bank. As players accumulate cubes, they exchange 10 cubes for 1 long. As they accumulate longs, they exchange 10 longs for 1 flat.

Salute!

The dealer gives one card to each of two players. Without looking at their cards, the players place them on their foreheads facing out. The dealer finds the sum of the numbers on the cards and says it aloud. Each player uses the sum and the number on the opposing player's forehead to find the number on his or her own card.

Beat the Calculator

One player is the Caller, who names two 1-digit numbers. Another player is the Brain, who adds the two numbers mentally. A third player is the Calculator, who adds the numbers with a calculator. The Brain tries to find the sum faster than the Calculator.

The sum is 12.

As You Help Your Child with Homework

As your child brings home assignments, you may want to go over the instructions together, clarifying them as necessary. The answers listed below will guide you through the Unit 6 Home Links.

Home Link 6-1

1. Answers vary.   2. Answers vary.

Home Link 6-2

1.

Quantity
29

Quantity	?
10	**Difference**

Rosa; Sample answer: $29 - 10 = ?$; \$19

2.

Quantity
15

Quantity	8
?	**Difference**

Sample answer: $8 + ? = 15$; 7 miles

Home Link 6-3

1. Sample answer: $16 + 7 = ?$; 23 inches
2. Sample answer: $24 + ? = 30$; 6 blocks

Home Link 6-4

1. 20 feet   2. 32 feet

Home Link 6-5

1. Sample answers: $11 + 6 - 8 = ?$; $11 + 6 = ?$
   and $17 - 8 = ?$; 9 children

Home Link 6-6

For 1–2, strategies will vary.

1. Sample estimate: $30 + 60 = 90$; 93
2. Sample estimate: $20 + 70 = 90$; 85
3. 246   4. 200; 70; 8
5. 350   6. 400; 20

Home Link 6-7

1. ||| . . . . |||| .

   $70 + 5 = 75$

2. || . . . . || . . . . .

   $40 + 12 = 52$

3. 532   4. 300; 40
5. 405   6. 600; 9

Home Link 6-8

In 1–3, sample estimates are shown.

1. $50 + 40 = 90$; 89
2. $30 + 80 = 110$; 108
3. $125 + 240 = 365$; 363

Home Link 6-9

1. 10

2. 8

3. a. 28   b. 25   c. 25   d. 29

Home Link 6-10

1. X X X X X X X
   X X X X X X X
   Sample answer: $8 + 8 = 16$

2. X X X X X
   X X X X X
   X X X X X
   X X X X X
   Sample answer: $6 + 6 + 6 + 6 = 24$

3. X X X X X X
   X X X X X X
   X X X X X X
   Sample answer: $7 + 7 + 7 = 21$

Making a Bar Graph

Family Note

Your child is exploring different ways to display data. One way to display data is in a bar graph. For the activity below, your child may have to ask a neighbor or call a relative to gather the needed pockets data.

Please return this Home Link to school tomorrow.

MRB
116

① Pick four people. Count the number of pockets on the clothes that each person is wearing. Record your data in the table.

Name	Number of Pockets

② Draw a bar graph for your data. First write each person's name on a line under the graph. Then color the bar above each name to show the number of pockets that each person has.

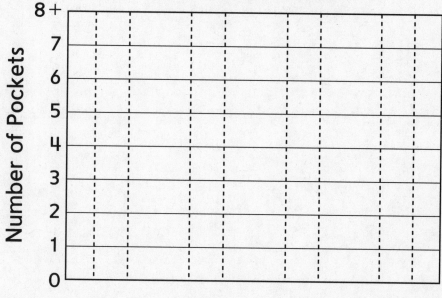

How Many Pockets?

Number of Pockets

8+
7
6
5
4
3
2
1
0

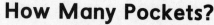

Names

Comparison Number Stories

Family Note

Today your child learned to use comparison diagrams. These diagrams help your child organize the information in a number story. When the information is organized, it is easier to decide whether to add or subtract to solve a problem.

Children use comparison diagrams to represent problems in which two quantities are compared. Sometimes children find the difference between the two quantities (as in Example 1 below). In other problems the difference is known, and children find one of the quantities (as in Example 2 below).

Quantity
49 fourth graders

Quantity
38 third graders

?

Difference

Example 1: There are 49 fourth graders and 38 third graders. How many more fourth graders are there than third graders?

Note that the number of fourth graders is being compared with the number of third graders.

• *Possible number models:* Children who think of the problem in terms of subtraction will write $49 - 38 = ?$ Other children may think of the problem in terms of addition: "Which number added to 38 will give me 49?" They will write the number model as $38 + ? = 49$.

• *Answer:* There are 11 more fourth graders than third graders.

Example 2: There are 53 second graders. There are 10 more second graders than first graders. How many first graders are there?

Note that the difference is known, and one of the two quantities is unknown.

Quantity
53

Quantity
?

10

Difference

• *Possible number models:* $53 - ? = 10$ or $10 + ? = 53$

• *Answer:* There are 43 first graders.

For Problems 1–2 on the next page, ask your child to explain the number models he or she wrote.

Please return the second page of this Home Link to school tomorrow.

Comparison Number Stories (continued)

For each number story, follow these steps:

MRB
30–31

- Write the numbers you know in the comparison diagram.
  Use ? for the number you need to find.

- Write a number model. Use ? for the number you don't know.

- Solve the problem and answer the question.

① Rosa has $29. Omeida has $10.

Who has more money? _____

How much more?

Number model:

_____

Rosa has $_____ more than Omeida.

| Quantity |
| Quantity |

Difference

② Omar ran 15 miles. Omar ran

8 more miles than Anthony.

How many miles did Anthony run?

Number model:

_____

Anthony ran _____ miles.

| Quantity |
| Quantity |

Difference

Addition and Subtraction Number Stories

Family Note

In today's lesson your child used diagrams to help solve addition and subtraction number stories. Diagrams help children organize the information from number stories, identify the missing information, and decide whether to add or subtract to solve the problem. Organizing information in a diagram also helps children write a number model using ? to represent what they don't know. Encourage your child to choose a diagram that best matches the way he or she sees the problem. There's no right or wrong diagram for a problem. What matters is that it matches the child's thinking.

Please return this Home Link to school tomorrow.

Do the following for each number story:

- Write a number model. Use ? to show what you need to find. To help, you may draw a

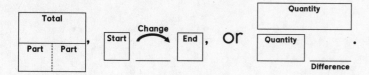

- Solve the problem and write the answer.

① It snowed 16 inches in Chicago on Friday night. It snowed 7 inches on Saturday night. How much snow did Chicago receive in all?

Number model: _____

Answer: _____ inches

② Evelyn has 30 blocks. She used 24 blocks to build a tower. How many blocks are not used for the tower?

Number model: _____

Answer: _____ blocks

Solving Problems

Family Note

In class today your child solved addition and subtraction number stories involving the heights and lengths of various animals. Some children used mental strategies to solve the stories. Others used tools such as base-10 blocks or open number lines. Others drew pictures or situation diagrams to help organize the information from the stories. Please do not teach your child a formal method, such as the addition method shown at the right. At this stage it is important for children to work with more concrete representations. Children will be introduced to a formal method for addition in Lessons 6-7 and 6-8.

$$\begin{array}{r} 52 \\ + 35 \\ \hline 87 \end{array}$$

Please return this Home Link to school tomorrow.

Solve the problems below. You may use base-10 shorthand, open number lines, or any other tool except a calculator to help. You may also draw pictures or diagrams.

(1) How tall are the ostrich and polar bear together?

9 feet

11 feet

Together they are _____ feet tall.

(2) How much longer is the giant squid than the crocodile?

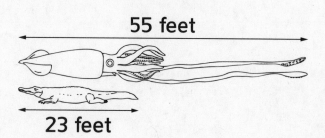

55 feet

23 feet

The giant squid is _____ feet longer than the crocodile.

Talk to someone about how you solved each problem.

Two-Step Number Stories

Family Note

In today's lesson your child solved two-step number stories, which can be broken into two parts and then solved in two steps. *For example:* Jonathan had 6 tickets for rides at the fair. His mother gave him 9 more. Then he gave 5 tickets to his friend. How many tickets does he have now?

To break this story into two parts, ask: What do you know from the story? (Jonathan had 6 tickets.) What happened first? (He received 9 more tickets.) What happened next? (He gave away 5 tickets.) What do you need to find out? (The number of tickets Jonathan has now.)

The first step is to figure out how many tickets Jonathan had after receiving some from his mother. The second step is to figure out how many tickets he had after giving some to his friend. Children are encouraged to solve two-step number stories using a variety of tools: drawings, open number lines, number grids, manipulatives, and diagrams.

They also learned to record either one or two number models for each number story—one for each part of the story or one number model to represent the whole story. *For example:* Use one number model, such as $6 + 9 - 5 = ?$, for both parts. Or, use two number models, such as $6 + 9 = ?$ and $15 - 5 = ?$, one for the first part and one for the second part. Answer: Jonathan now has 10 tickets.

Ask your child to explain the steps he or she takes to solve the problem below. Discuss how his or her number model(s) relates to the number story.

Please return this Home Link to school tomorrow.

- Write a number model or number models. Use ? to show the number you need to find.
  To help, you may draw a

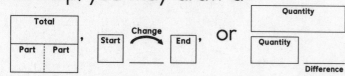

- Solve the problem and write the answer.

① At the beach, 11 children were playing in the sand. Then 6 more children joined them. Then 8 decided to go swimming. How many children were still playing in the sand?

Unit
children

Number model(s): _____

Answer: _____ children

Addition Strategies

Family Note

Everyday Mathematics encourages children to use a variety of strategies to solve computation problems. Doing so helps children develop a sense for numbers and operations, rather than simply memorizing a series of steps.

We suggest that you give your child an opportunity to explore and choose addition strategies that he or she feels comfortable using. At some point you may want to share the method that you know from your own school experience. However, please allow your child some time to use his or her own methods before doing so.

Below are three examples of methods that your child might use to solve 2-digit addition problems.

Counting Up

$47 + 33 = ?$ ⟵ "My problem"

47, 57, 67, 77 ⟵ "Start at 47. Count up 30 by 10s."

78, 79, 80 ⟵ "Count 3 more."

80 ⟵ "The answer is 80."

Combining 10s and 1s

$29 + 37 = ?$ ⟵ "My problem"

$20 + 30 = 50$ ⟵ "Add the 10s."

$9 + 7 = 16$ ⟵ "Add the 1s."

$50 + 16 = 66$ ⟵ "Put the 10s and 1s together. The answer is 66."

Making Friendly Numbers

$52 + 29 = ?$ ⟵ "My problem"

30 ⟵ "30 is close to 29. Just add 1 more to get 30."

$52 + 30 = 82$ ⟵ "52 plus 30 is 82."

$82 - 1 = 81$ ⟵ "Take away 1 because I added 1 to get 30. The answer is 81."

Encourage your child to use a ballpark estimate as a way to check whether an answer to a computation problem makes sense. *For example:* In $29 + 37$, 29 is close to 30 and 37 is close to 40. Because $30 + 40 = 70$, a ballpark estimate is 70. The final answer of 66 is close to 70, so 66 is a reasonable answer. Your child can make a ballpark estimate before or after solving the problem.

Please return the second page of this Home Link to school tomorrow.

Addition Strategies (continued)

For each problem:

- Make a ballpark estimate.

- Solve the problem using any strategy you choose. Use words or pictures to show your thinking.

- Check to make sure your answer makes sense.

Unit

① 34 + 59 = ?

Ballpark estimate:

_____

Strategy:

② 17 + 68 = ?

Ballpark estimate:

_____

Strategy:

34 + 59 = _____

17 + 68 = _____

Choose one of the problems above. Explain your estimate to someone at home. Then explain how you checked to make sure your answer made sense.

Practice

Complete each number sentence to show the expanded form.

③ _____ = 200 + 40 + 6   ④ 278 = _____ + _____ + _____

⑤ 300 + 50 = _____   ⑥ 420 = _____ + _____

Adding with Base-10 Blocks

Family Note

Today children used base-10 blocks to help them add numbers. Three types of base-10 blocks were used: A cube represents 1. A long (a rod that is 10 cubes long) represents 10. A flat (a square that is 10 cubes long and 10 cubes wide) represents 100.

To solve 24 + 32 with base-10 blocks, children first represent each number with blocks or base-10 shorthand:

$$\begin{array}{r} 24 \\ + \ 32 \end{array} \quad \begin{array}{l} | | \ \bullet \bullet \bullet \bullet \\ | | | \ \bullet \bullet \end{array}$$

Then children combine the blocks according to type (longs with longs; cubes with cubes) and count each type of block: 5 longs show 5 tens, or 50; 6 cubes show 6 ones, or 6. The 50 and the 6 are called *partial sums* because they are parts of the final sum. Finally, children add the partial sums to find the total: 50 + 6 = 56.

Children also use base-10 blocks to add 3-digit numbers by adding the 100s, 10s, and 1s separately and then combining the partial sums to find the total.

Please return this Home Link to school tomorrow.

Use base-10 shorthand to show each number. Then write the partial sums and find the total sum.

Unit

MRB
71, 76

①
$$\begin{array}{r} 34 \\ + \ 41 \end{array}$$

②
$$\begin{array}{r} 27 \\ + \ 25 \end{array}$$

_____ + _____ = _____   _____ + _____ = _____

Explain to someone at home how you use base-10 blocks to add.

Practice

Complete each number sentence to show the expanded form of a number.

③ _____ = 500 + 30 + 2

④ 340 = _____ + _____

⑤ 400 + 5 = _____

⑥ 609 = _____ + _____

More Partial Sums

Family Note

In the previous lesson your child used base-10 blocks to help find partial sums. Today your child used expanded form. Expanded form shows numbers broken apart into a sum of place-value pieces, such as hundreds, tens, and ones. For example, the expanded form for 324 is 300 + 20 + 4.

To solve 324 + 255, your child can first write or think about each number in expanded form, then use the expanded form to help find the partial sums:

Think:
$300 + 200 =$
$20 + 50 =$
$4 + 5 =$

```
    324
  + 255
    500
     70
      9
    579
```

Think:
$300 + 20 + 4$
$200 + 50 + 5$

Encourage your child to use place-value language when working with this method. For example, when adding the 100s in this example, guide your child to say "300 + 200 = 500," not "3 + 2 = 5." Writing the expanded form can help children remember to use the correct language.

This method of finding partial sums and then combining the partial sums to find the total is called partial-sums addition. Partial-sums addition was introduced only recently, so allow plenty of time for practice before expecting your child to use it easily.

Please return this Home Link to school tomorrow.

Fill in the unit box. For each problem:

MRB
80

• Make a ballpark estimate. Solve the problem using partial-sums addition. Show your work.

• Use your ballpark estimate to check if your answer makes sense.

Unit

① Ballpark estimate: ② Ballpark estimate: ③ Ballpark estimate:

_____

```
   53
 + 36
```

_____

```
   27
 + 81
```

```
  126
+ 237
```

Subtraction Number Stories

Family Note

In today's lesson, your child solved subtraction number stories using different tools and strategies based on place-value concepts and explained his or her thinking in drawings and words. Being able to solve problems in multiple ways and explain their strategies helps children become flexible problem solvers.

As your child solves these problems, ask him or her to explain the strategy.

Please return this Home Link to school tomorrow.

① Sam is on a baseball team. This year he set a goal of scoring 36 runs for his team. So far Sam has scored 26 runs. How many more runs does Sam need to score in order to meet his goal?

MRB
30-31

_____ runs

② Sam helped his mother unload the dishwasher. As he was putting the silverware away, Sam counted 21 spoons and 13 forks. How many more spoons than forks did Sam unload?

_____ spoons

Practice

Unit

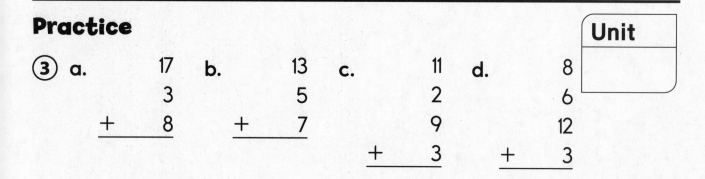

③ a.
```
    17
     3
+    8
```

b.
```
    13
     5
+    7
```

c.
```
    11
     2
     9
+    3
```

d.
```
     8
     6
    12
+    3
```

How Many?

Family Note

Your child has been working with arrays to develop readiness for multiplication. Arrays are rectangular arrangements of objects that have the same number of objects in each row. For example, a 3-by-5 array is shown at the right.

X X X X X
X X X X X
X X X X X

Your child found the total number of objects in each array and learned to write addition number models to represent arrays. One example of an addition number model for this array is $5 + 5 + 5 = 15$. There are 15 Xs in all.

When your child writes an addition number model to show the number of objects in a 5-by-4 array, he or she is building understanding of the meaning of four 5s, or 4×5.

Please return this Home Link to school tomorrow.

MRB
32-33

① Draw an array with 2 rows of Xs with 8 Xs in each row.

Write an addition number model for the array.

_____

② Draw an array with 4 rows of Xs with 6 Xs in each row.

Write an addition number model for the array.

_____

③ Draw an array with 3 rows of Xs with 7 Xs in each row.

Write an addition number model for the array.

_____

Whole Number Operations and Measurement and Data

In Unit 7 children revisit combinations of 10 and answer questions like: "What must I add to 4 to get to 10?" They extend this idea to larger numbers and answer questions like: "What must I add to 47 to get to 50?" and "What must I add to 28 to get to 40?"

I need to add a number to 28 to get to 40. What number, added to 8, will give me 10? It's 2, so 28 + 2 = 30. What number, added to 30, will give me 40? It's 10, because 30 + 10 = 40. Finally, 2 + 10 = 12, so I have to add 12 to get to 40.

Children also discuss strategies for solving addition problems that have more than two addends, such as 14 + 2 + 6 + 12.

In later lessons in this unit, children use two length units—meters and yards—to measure longer lengths and distances, and they develop personal references for these units to use when estimating lengths. Children also collect real-life data and display it in tables and graphs. For example, children collect data by measuring the lengths of their standing jumps. Then they display their data on a line plot.

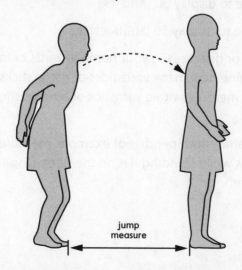

jump measure

Please keep this Family Letter for reference as your child works through Unit 7.

Vocabulary   Important terms in Unit 7:

multiple of 10 A product of 10 and a counting number. The multiples of 10 are 10, 20, 30, 40, and so on.

personal reference A convenient approximation for a standard unit of measurement. *For example:* For many people the distance from the tip of the thumb to the first joint is approximately 1 inch.

yard A U.S. customary unit of length equal to 3 feet, or 36 inches.

meter The basic metric unit of length from which other metric units of length are derived. One meter is equal to 100 centimeters, or 1,000 millimeters.

arm span The distance from fingertip to fingertip of outstretched arms.

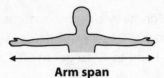

Arm span

line plot A sketch of data in which check marks, Xs, or other symbols above a labeled line show the frequency of each value.

```
              x
              x    x
              x    x
    x         x    x
    x         x    x              x
    0    1    2    3    4
      Number of Siblings
        A line plot
```

Do-Anytime Activities

To work with your child on the concepts taught in this unit and previous units, try these interesting and rewarding activities:

1. If you have a calculator at home, practice making multiples of 10 from given numbers or breaking apart multiples of 10. *For example:*

   • Enter 33. What needs to be done to display 50? (Add 17.)

   • Enter 70. What needs to be done to display 62? (Subtract 8.)

   • Enter 57. What needs to be done to display 90 (Add 33.)

   • Enter 78. What needs to be done to display 50 (Subtract 28.)

2. Ask your child to estimate lengths or distances in your home in yards or in meters. To estimate, ask your child to imagine how many yardsticks or metersticks might fit along a length or a distance. Then measure with a yardstick or a meterstick to check the estimates.

3. Collect a simple set of data from family and friends. For example, measure how high they can reach with their fingertips while standing flat on the floor. Display the data in a tally chart, on a line plot, or both.

Building Skills through Games

In Unit 7 your child will practice mathematical skills by playing the following games:

Hit the Target

Players choose a 2-digit multiple of 10 (such as 10, 20, or 30) as a target number. One player chooses a starting number less than or larger than the target number, which the second player enters into a calculator. The second player tries to change it to the target number by adding or subtracting numbers on the calculator.

Basketball Addition

This game is played by two teams of three to five players each. Players score points by rolling a 20-sided die and recording the number (or rolling three 6-sided dice and recording the sum). The team score is determined by adding the scores of all the players on each team. The team that scores more points than the other wins the game.

Beat the Calculator

One player is the Caller, who names two 1-digit numbers. Another player is the Brain, who adds the two numbers mentally. A third player is the Calculator, who adds the numbers with a calculator. The Brain tries to find the sum faster than the Calculator.

Addition/Subtraction Spin

Players spin a spinner to determine a 3-digit number. Then they roll a die to see if they should add 10 or 100 to the 3-digit number or subtract 10 or 100 from it. Players do the computation mentally.

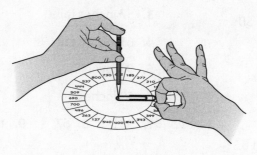

As You Help Your Child with Homework

When your child brings home assignments, you may want to go over the instructions together, clarifying them as necessary. The answers listed below will guide you through the Unit 7 Home Links.

Home Link 7-1

1. 6; 7; 5; 9; 2

2. 6; 7; 5; 9; 8

3.

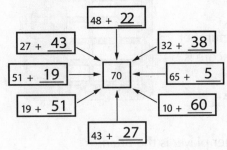

Home Link 7-2

1. Sample model: 13 + 7 + 6 = 26

2. Sample model: 8 + 22 + 5 = 35

3. Sample model: 25 + 15 + 9 = 49

4. Sample model: 29 + 11 + 6 + 4 = 50

5. 69   **6.** 70   **7.** 62

8. 83   **9.** 169   **10.** 204

Home Link 7-3

1. 35; 25; Team A   **2.** 30; 35; Team B

3. 29; 40; Team B   **4.** 45; 59; Team B

Home Link 7-4

1–3. Answers vary.

4. 94   **5.** 67   **6.** 34   **7.** 54

Home Link 7-5

1. Answers vary.   **2.** Answers vary.

3. More centimeters; Sample answer: Centimeters are shorter, so it takes more of them to measure the same height.

4. 2   **5.** 50   **6.** 93   **7.** 41

Home Link 7-6

1–4. Answers vary.

5. 60   **6.** 75   **7.** 43   **8.** 8

Home Link 7-7

1. 57, 60, 62, 64, 64, 68, 71, 72

2. 57 inches   **3.** 72 inches   **4.** 15 inches

5. 98   **6.** 29

Home Link 7-8

1. 2 players

2. 0 players

3. 57 inches tall

4. 63 inches tall

5. 9 players

6. 59 inches

7. 39   **8.** 67   **9.** 19   **10.** 61

Home Link 7-9

Favorite Vegetables Picture Graph

Carrots	Peas	Corn	Other
			☺
	☺		☺
☺	☺		☺
☺	☺	☺	☺
☺	☺	☺	☺
☺	☺	☺	☺

Name of Vegetable

KEY: Each ☺ = 1 child

1. 26   **2.** 67   **3.** 2   **4.** 42

Missing Addends

Family Note

In this lesson your child used mental strategies to find differences between 2-digit numbers and larger multiples of 10. For example, your child found what number added to 44 equals 50. (The answer is 6.) In Problems 1–2 your child will find the difference between a number and the next-larger multiple of 10. In Problem 3 your child will find different combinations of numbers that add to 70. If your child has difficulty with this problem, suggest first adding 1s to the first number in each combination to find the next-larger multiple of 10. For example, add 2 to 48 to make 50. Then add 20 (or two 10s) to 50 to make 70. Finally, add 2 + 20 to find the answer, 22. So 48 + 22 = 70.

Please return this Home Link to school tomorrow.

① $4 +$ _____ $= 10$

　$10 = 3 +$ _____

　_____ $+ 5 = 10$

　$10 =$ _____ $+ 1$

　$8 +$ _____ $= 10$

② $54 +$ _____ $= 60$

　$90 = 83 +$ _____

　$75 +$ _____ $= 80$

　$40 = 31 +$ _____

　_____ $+ 42 = 50$

③ Make 70s. Show someone at home how you did it.

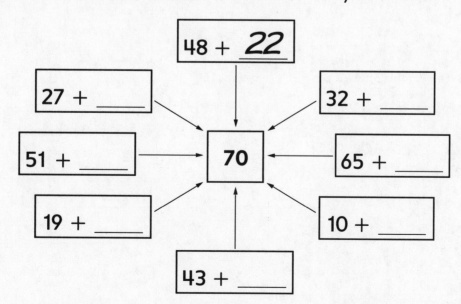

Adding Three or More Numbers

Family Note

Today your child added more than 2 addends. Changing the order of the addends can make it easier to find the sum. For example, when adding 17, 19, and 23, some people may first calculate 17 + 23, which equals 40, and then add 19 (40 + 19 = 59). For Problems 1–4, help your child look for easy combinations. Before working on Problems 5–10, you might go over the example with your child.

Please return this Home Link to school tomorrow.

For each problem:

- Think about an easy way to add the numbers.

- Write a number model to show the order in which you are adding the numbers.

- Find each sum. Tell someone at home why you added the numbers in that order.

MRB
44

①

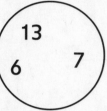

13

6   7

Number model:

___ + ___ + ___ = ___

②

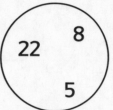

8

22

5

Number model:

___ + ___ + ___ = ___

③

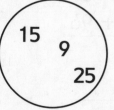

15

9

25

Number model:

___ + ___ + ___ = ___

④

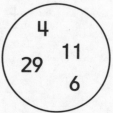

4

11

29

6

Number model:

__ + __ + __ + __ = __

Adding Three or More Numbers (continued)

Add. Use the partial-sums method.

Example:

$$
\begin{array}{r} 33 \\ 42 \\ +\ 11 \\ \end{array}
$$

Add the tens. $\rightarrow (30 + 40 + 10)$ $\rightarrow$ 80

Add the ones. $\rightarrow (3 + 2 + 1)$ $\rightarrow$ 6

Add the partial sums. $\rightarrow (80 + 6)$ $\rightarrow$ 86

Practice

(5)
$$
\begin{array}{r} 23 \\ 32 \\ +\ 14 \\ \hline \end{array}
$$

(6)
$$
\begin{array}{r} 14 \\ 29 \\ +\ 27 \\ \hline \end{array}
$$

(7)
$$
\begin{array}{r} 8 \\ 19 \\ +\ 35 \\ \hline \end{array}
$$

(8)
$$
\begin{array}{r} 46 \\ 25 \\ +\ 12 \\ \hline \end{array}
$$

(9)
$$
\begin{array}{r} 21 \\ 40 \\ 45 \\ +\ 63 \\ \hline \end{array}
$$

(10)
$$
\begin{array}{r} 14 \\ 9 \\ 85 \\ +\ 96 \\ \hline \end{array}
$$

Who Scored More Points?

Family Note

In this lesson your child added three or more 1-digit and 2-digit numbers. As your child completes the problems below, encourage him or her to share the different ways in which the points can be added. Your child might add all the 10s first and then add all the 1s. For example, $20 + 5 + 4 + 6 = 20 + 15 = 35$. Your child may also look for combinations of numbers that are easier to add. In Game 1, for example, first add 14 and 6 to get 20 and then add 15 to get 35.

Please return this Home Link to school tomorrow.

Do the following for each problem:

Unit
points

- Add the points for each team.

- Decide which team scored more points.
  The team with more points wins the game.

- Circle your answer.

① Game 1

Team A:

$15 + 14 + 6 =$ _____

Team B:

$5 + 13 + 7 =$ _____

Who won? A or B

② Game 2

Team A:

$12 + 6 + 4 + 8 =$ _____

Team B:

$5 + 10 + 19 + 1 =$ _____

Who won? A or B

③ Game 3

Team A:

$17 + 4 + 5 + 3 =$ _____

Team B:

$2 + 11 + 9 + 18 =$ _____

Who won? A or B

④ Game 4

Team A:

$7 + 4 + 16 + 13 + 5 =$ _____

Team B:

$22 + 9 + 8 + 3 + 17 =$ _____

Who won? A or B

Using Measurement

Family Note

In class today your child measured distances with a yardstick. Talk with your child about measurements you use at your job, around the house, in sports, or in other activities. If you don't have measuring tools to show your child, you might find pictures of measuring tools online or in a catalog, magazine, or book. Discuss with your child how these tools are used.

Please return this Home Link to school tomorrow.

① Talk with people at home about how they use measurements at home, at their jobs, or in other activities.

② Ask people at home to show you the tools they use for measuring. Write the names of some of these tools. Be ready to talk about your list in class.

_____   _____

_____   _____

_____   _____

③ Look for measurements in pictures, in newspapers, or magazines. For example, an ad might tell the height of a bookcase or how much a container holds. Ask an adult if you may bring the pictures to school for our Measures All Around Museum. Circle the measurements.

Practice

Solve.

Unit

④ $93 + 1 =$ _____

⑤ _____ $= 6 + 61$

⑥ _____ $= 26 + 8$

⑦ $5 + 49 =$ _____

Measuring Height

Family Note

In this lesson your child was introduced to a metric unit of length called the meter. One meter is equal to 100 centimeters. We compared metersticks to yardsticks and noticed that 1 meter is a little longer than 1 yard. Then we used tools such as rulers, yardsticks, metersticks, and tape measures to measure lengths. Your child may wonder why there are two standard units—yards and meters—that are nearly the same size. You may want to explore this issue by searching online for information about the metric and U.S. customary systems of measurement.

If you don't have tools to measure length with metric units at home, you and your child can cut pieces of string or strips of paper to match the heights of a table and an adult. Your child can bring the string or strips to school to measure.

Please return this Home Link to school tomorrow.

① Work with someone at home to measure the height of a table.

The table is about _____ centimeters high.

The table is about _____ meters high.

② Measure the height of an adult.

The adult is about _____ centimeters tall.

The adult is about _____ meters tall.

③ Are there more centimeters or more meters in your measurements? _____ Explain.

_____

_____

Practice

Unit

④ $18 +$ _____ $= 20$

⑤ _____ $+ 3 = 53$

⑥ _____ $= 86 + 7$

⑦ $8 + 33 =$ _____

Comparing Arm Spans

Family Note

In today's lesson your child measured his or her standing jump and arm span in both centimeters and inches. Help your child compare his or her arm span to someone else's arm span at home. Also help your child find objects around the house that are about the same length as his or her arm span.

Please return this Home Link to school tomorrow.

My arm span is about _____ inches long.

1. Tell someone at home about how long your arm span is in inches.

2. Compare your arm span to the arm span of someone at home. Can you find someone who has a longer arm span than you do? Is there someone at home who has a shorter arm span?

   _____ has a longer arm span than I have.

   _____ has a shorter arm span than I have.

3. List some objects that are about the same length as your arm span.

   _____ _____ _____

4. Explain how you know the objects you listed in Problem 3 are about the same length as your arm span.

   _____

   _____

Practice

Solve.

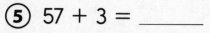

5. $57 + 3 =$ _____

6. $4 + 71 =$ _____

7. _____ $= 34 + 9$

8. $48 +$ _____ $= 56$

Unit

Interpreting Data

Family Note

In this lesson your child examined classroom data on the length of classmates' standing jumps. The class found the shortest jump length and the longest jump length and calculated the difference between the lengths. They also made a line plot based on the data.

Please return this Home Link to school tomorrow.

The track team collected these standing-jump data:

Jumper	Standing-Jump Length
Fran	68 inches
Arturo	72 inches
Louise	57 inches
Kelsey	71 inches
Keisha	60 inches
Ray	64 inches
Maria	64 inches
Ben	62 inches

① List the inches for each jump in order from shortest to longest.

_____, _____, _____, _____, _____, _____, _____, _____

② What is the shortest jump length? _____ inches

③ What is the longest jump length? _____ inches

④ What is the difference between the longest jump length and the shortest jump length? _____ inches

Practice

⑤ _____ = 1 + 97

⑥ 23 + 6 = _____

Interpreting Data

Family Note

Today your child represented class arm span data in a frequency table and on a line plot. Line plots like the one below help us organize and display data. Each X in this line plot represents one basketball player. Help your child use the data in the line plot to answer the questions.

Please return this Home Link to school tomorrow.

Ms. Ortiz is a basketball coach. She measured the height of each player on the team. Then she made this line plot.

Players' Heights

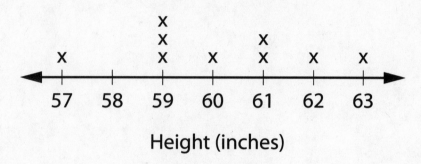

Height (inches)

1. How many players are 61 inches tall? _____ players

2. How many players are 58 inches tall? _____ players

3. The shortest player is _____ inches tall.

4. The tallest player is _____ inches tall.

5. How many players did Ms. Ortiz measure? _____ players

6. Which height occurs most often? _____ inches

Practice

Unit

7. $33 + 6 =$ _____

8. _____ $= 65 + 2$

9. _____ $+ 3 = 22$

10. $9 + 52 =$ _____

Vegetable Picture Graph

Family Note

Today your child drew a picture graph, which uses pictures or symbols to show data. The key on a picture graph tells what each picture is for. Have your child use the data table to draw the graph.

Please return this Home Link to school tomorrow.

Favorite Vegetables

Name of Vegetable	Number of People
Carrots	4
Peas	5
Corn	3
Other	6

Favorite Vegetables Picture Graph

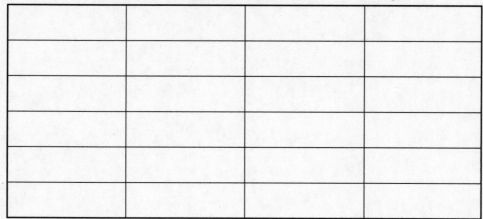

_____ _____ _____ _____

Name of Vegetable

KEY: ☺ = 1 child

Practice

① _____ = 21 + 5   ② 63 + 4 = _____

③ _____ + 88 = 90   ④ 7 + 35 = _____

Unit 8: Family Letter

Geometry and Arrays

In Unit 8 children explore 2-dimensional shapes, including triangles, quadrilaterals, pentagons, and hexagons. They describe and sort the shapes according to their attributes, such as number of sides, length of sides, number of angles, and whether they have right angles or parallel sides.

 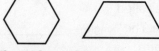

These shapes each have at least one right angle.   **These shapes have no right angles.**

Children also look for 2-dimensional shapes in 3-dimensional objects. For example, they look at a cube and notice that each face, or side, of the cube is a square.

After these shape activities, children also explore techniques for partitioning rectangles into rows and columns of same-size squares. These activities lay the foundation for area measurement in Grade 3.

This rectangle is partitioned into 2 rows and 6 columns of squares.

In the last part of the unit, children solve number stories involving equal groups of objects. In some cases equal groups are small clusters of objects, such as petals on flowers. In other cases the equal groups are the rows or columns of rectangular arrays.

Equal groups of petals: 3 flowers with 5 petals on each flower is 15 petals in all.   **An array of chairs: 3 rows with 5 chairs in each row is 15 chairs in all.**

Children build equal groups and arrays with counters and explore strategies for finding how many counters there are in all. These activities lay the foundation for work with multiplication in Grade 3.

Please keep this Family Letter for reference as your child works through Unit 8.

Vocabulary Important terms in Unit 8:

attribute (of a shape) A feature of a shape or a common feature of a set of shapes. Examples of shape attributes include the number of sides and the number of right angles.

angle Two rays or two line segments with a common endpoint. The rays or segments are called the *sides* of the angle. The sides of a polygon form angles at each corner, or vertex, of the polygon.

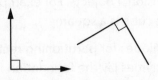

right angle A 90-degree angle. Also known as a square corner.

parallel lines Two lines in a plane are parallel if they never intersect or cross. Two parallel lines are always the same distance apart. Two line segments in a plane are parallel if they can be extended to form parallel lines. If two sides of a polygon are parallel line segments, that polygon has a pair of parallel sides.

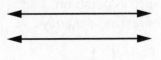

Parallel lines

polygon A 2-dimensional figure formed by three or more line segments (sides) that meet only at their endpoints to make a closed path. The sides may not cross one another.

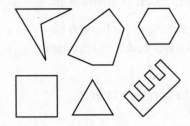

cube A 3-dimensional shape with exactly 6 square faces.

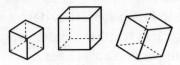

face In *Everyday Mathematics*, a flat surface on a 3-dimensional shape.

row A horizontal arrangement of objects or numbers in an array or a table.

column A vertical arrangement of objects or numbers in an array or a table.

partition To divide a shape into smaller shapes. In *Second Grade Everyday Mathematics*, children partition rectangles into rows and columns of same-size squares. See the example on the first page of this letter.

equal groups Sets with the same number of elements, such as cars with 5 passengers each or boxes containing 100 paper clips each.

array An arrangement of objects in a regular pattern. In *Second Grade Everyday Mathematics*, children work with rectangular arrays, which are arrangements of objects in rows and columns that form rectangles. All rows have the same number of objects, and all columns have the same number of objects. The rows and columns in a rectangular array are one way of representing equal groups.

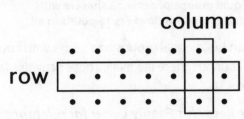

Do-Anytime Activities

To work with your child on the concepts taught in this unit and previous units, try these interesting and rewarding activities:

1. Point to everyday objects and ask your child to identify the shapes he or she sees and describe their attributes. For example, your child might see rectangles on the sides of a shoe box and point out the parallel sides and right angles, or he or she might see hexagons on a soccer ball and note that they each have 6 equal-length sides.

2. Name a shape (such as a rectangle) or an attribute (such as a right angle) and ask your child to find an object with that shape or attribute. For example, if asked to find a shape with 4 right angles, your child might identify a book cover or a doorway.

3. Look for real-life examples of equal groups or arrays and ask your child to figure out how many objects there are in each one. For example, most telephone keypads have 4 rows of 3 keys each. That's $3 + 3 + 3 + 3 = 12$ (or $4 + 4 + 4 = 12$) keys in all. Other examples of real-life equal groups or arrays might include floor or ceiling tiles, window panes, or packages of pencils or markers.

Building Skills through Games

In Unit 8 your child will practice mathematical skills by playing a variety of games, including the following new games.

Shape Capture

Players have a set of Shape Cards spread out in front of them. One at a time, players draw an Attribute Card and "capture" all the shapes that have that attribute. The player who captures the most shapes wins.

The Number-Grid Difference Game

Each player draws two number cards and uses them to form a 2-digit number. Players mark the numbers on a number grid, and one player finds the difference between the two numbers. The difference is that player's score for the round.

Array Concentration

Players arrange a set of *Array Concentration* Number Cards and Array Cards facedown in front of them. A player flips over one of each type of card. If the cards "match"—that is, if the number on the number card equals the total number of dots in the array—the player takes the cards and takes another turn.

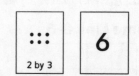

These cards match because there are 6 dots in the array.

Array Bingo

Players arrange a set of array cards to form a bingo card. Children take turns drawing number cards and calling out the number. If players have an array on their bingo card that has that number of dots, they turn over the array card. The first player to turn over three arrays in a row (vertically, horizontally, or diagonally) wins.

As You Help Your Child with Homework

As your child brings home assignments, you may go over the instructions together, clarifying them as needed. The answers below will guide you through the Unit 8 Home Links.

Home Link 8-1
1.–4. Answers vary.
5. 36   **6.** 52   **7.** 83

Home Link 8-2
1.   2.

3.

4.

Home Link 8-3
1.–3. Sample drawings are given.

1. ; Hexagon   **2.** ; Pentagon

3. ; Sample answer: Quadrilateral

4. Yes. Sample answer: They are all closed shapes with straight sides that don't cross.

Home Link 8-4
1. Answers vary.   **2.** Answers vary.

3. Sample answer: A triangle has 3 sides, and a quadrilateral has 4 sides.

4. Answers vary.

5a. 61   **5b.** 73   **5c.** 94   **5d.** 72

Home Link 8-5
1.–3. Answers vary.   **4.** 45   **5.** 50   **6.** 94

Home Link 8-6

1. 12   **2.** 12

Home Link 8-7

1.

6 squares

2. 39   **3.** 80   **4.** 96

Home Link 8-8
1. 10 fingers; Sample answers: $5 + 5 = 10$; $2 \times 5 = 10$

2. 12 muffin cups; Sample answers: $4 + 4 + 4 = 12$; $3 \times 4 = 12$

3. 8 shoes; Sample answers: $2 + 2 + 2 + 2 = 8$; $2 \times 4 = 8$

Home Link 8-9
1. 12; Sample answers: $3 + 3 + 3 + 3 = 12$; $4 \times 3 = 12$

2. 15; Sample answers: $5 + 5 + 5 = 15$; $3 \times 5 = 15$

3. 8; Sample answers: $2 + 2 + 2 + 2 = 8$; $4 \times 2 = 8$

4. 55   **5.** 91   **6.** 94

Home Link 8-10
1. 20; Sample answers: $4 + 4 + 4 + 4 + 4 = 20$; $5 + 5 + 5 + 5 = 20$; $4 \times 5 = 20$; $5 \times 4 = 20$

2. 2; Sample answers: $1 + 1 = 2$; $1 \times 2 = 2$; $2 \times 1 = 2$

3. 8; Sample answers: $4 + 4 = 8$; $2 + 2 + 2 + 2 = 8$; $2 \times 4 = 8$; $4 \times 2 = 8$

Home Link 8-11
1. Answers vary.

Shapes

Family Note

In this lesson children examined different shapes, such as triangles, quadrilaterals, pentagons, and hexagons. They also discussed these shapes' attributes—or characteristics—such as the number of sides, the number of angles, whether the sides are parallel, and whether the angles are right angles.

Look at the various shapes shown below. Examples of these shapes can be found in objects you see every day, such as yield signs (which resemble triangles) or TV screens (quadrilaterals). As your child cuts out pictures of shapes, discuss each one. Count the number of sides and angles and try to name the shapes. Talk about how they are alike and how they are different.

Please return this Home Link to school tomorrow or as requested by the teacher.

① Cut out pictures from newspapers and magazines that show 3-sided, 4-sided, 5-sided, and 6-sided shapes. Ask an adult for permission first.

② Glue or tape each picture to a sheet of paper.

③ Label some of the pictures with their shape names.

④ Bring your pictures to school.

Triangles	Quadrilaterals
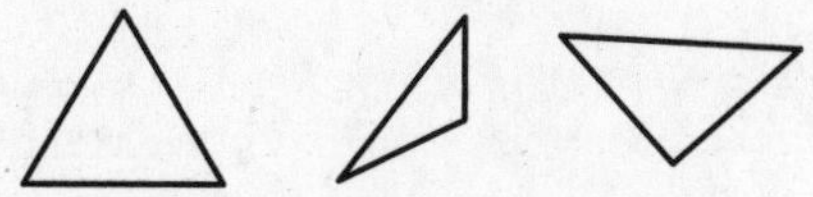	
Pentagons	Hexagons
	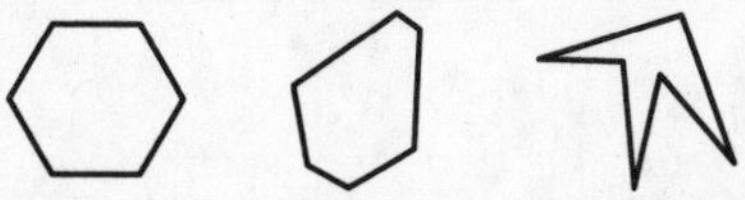

Practice

Add.

Unit

⑤ 24 + 12 = _____   ⑥ 33 + 19 = _____   ⑦ 47 + 36 = _____

one hundred seventy-one   171

Attributes of Shapes

Family Note

In this lesson children played a game called *Shape Capture*, in which they "captured" shapes based on attributes: the number of sides, the number of angles and vertices, the number of right angles, the number of pairs of parallel sides, and side lengths. After your child has completed the Home Link, discuss how the shapes he or she circled are different from the others.

Please return this Home Link to school tomorrow.

① Look at the number of right angles. Circle the shape(s) with 1 right angle.

② Look at the number of sides and angles. Circle the shape(s) with 5 sides and 5 angles and 5 vertices.

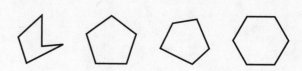

③ Look at the lengths of the sides. Circle the quadrilateral(s) with 2 pairs of equal-length sides.

④ Look at the opposite sides. Circle the quadrilateral(s) with one or more pairs of parallel sides.

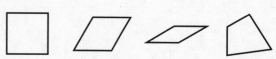

Shapes Museum

Family Note

In today's lesson children used straws and twist ties to build polygons and then drew the shapes. Children learned that polygons are closed figures made up of all straight sides that do no cross.

Polygons: NOT polygons:

Please return the top part of this Home Link to school tomorrow.

For Problems 1–3, draw the polygon and write its name on the line.

(1) 6-sided polygon: (2) 5-sided polygon: (3) 4-sided polygon:

_____   _____   _____

(4) Are these three shapes all polygons? Explain.

_____

_____

Shapes Museum

For the next few days our class will collect items to put into a Shapes Museum. Starting tomorrow, bring items such as boxes, soup cans, party hats, pyramids, and balls to school. Ask an adult for permission to bring in these items.

Drawing Shapes

Family Note

In this lesson your child learned about attributes of quadrilaterals (four-sided figures). We drew quadrilaterals with certain numbers of right angles and wrote about how we knew the shape had the correct attributes. We will revisit how to recognize and draw quadrilaterals and other shapes for the rest of the school year.

Please return this Home Link to school tomorrow.

① Draw a quadrilateral that has four right angles.
Use the dots to help you.

② Draw a triangle that has one right angle.
Use the dots to help you.

③ Name something that is different about a quadrilateral and a triangle.

_____

_____

④ Show someone at home how you can test if an angle is a right angle.

Practice

⑤ a. 23
     + 38

   b. 56
     + 17

   c. 26
     + 68

   d. 36
     + 36

3-Dimensional Shapes

Family Note

In this lesson children described and compared different 3-dimensional shapes. The class also created a Shapes Museum using the objects children brought to school. Read the list of shapes below with your child. Together, find examples of the shapes.

Please return this Home Link to school tomorrow.

Work with someone to make a list of things that have these shapes.

MRB
134–136

① Cube

② Rectangular prism

③ Cylinder

_____   _____   _____

_____   _____   _____

_____   _____   _____

Practice

Add.

Unit

④　　21
　　+ 24

⑤　36 + 14 = _____

⑥　　38
　　+ 56

Partitioning Rectangles

Family Note

In class today your child learned how to use same-size shapes to partition, or divide, a shape into smaller shapes. Understanding how to partition shapes helps lay the foundation for area measurement in Grade 3. Help your child cut out the squares below and use them to completely cover Rectangle A on the next page without gaps or overlaps. After the squares are positioned, your child can glue or tape them in place. Then your child will draw lines on Rectangle B to show how the squares are arranged on Rectangle A. He or she can use one of the extra squares to help with the partitioning. Do not expect the squares your child draws to be exactly the same size. The goal is for your child to draw the correct number of squares arranged in rows and columns.

Please return the second page of this Home Link to school tomorrow.

- Carefully cut out the small squares below.

- Use the squares to completely cover Rectangle A on the next page without any overlaps or gaps. You will not need them all.

- Glue or tape the squares in place.

- Draw lines on Rectangle B to show where you put the squares on Rectangle A.

- Answer the questions below the rectangles.

Family Note

- Carefully cut out the small squares below.

- Use the squares to completely cover Rectangle A on the next page without any overlaps or gaps. You will need the glue or tape the squares in place.

- Draw lines on Rectangle A to show where you put the squares on Rectangle A.

- Answer the questions below the rectangles.

Partitioning Rectangles (continued)

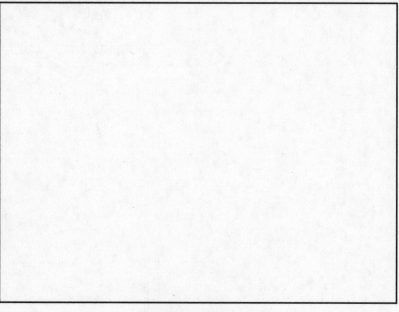

Rectangle A

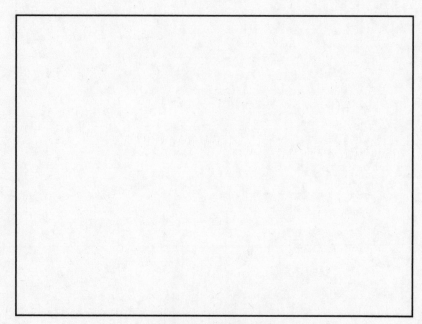

Rectangle B

① How many squares did you use to cover Rectangle A? _____

② How many squares did you draw on Rectangle B? _____

More Partitioning Rectangles

Family Note

In this lesson children continued their work partitioning rectangles into same-size squares. They used a square block and then a picture of a square to help them determine the size of the squares needed to cover their rectangles. Finally, they partitioned rectangles into given numbers of rows with a specific number of squares in each row. Children are not expected to draw perfect rows of squares. The goal of the activity below is for them to make rows that are close to the same height and squares that are about the same size.

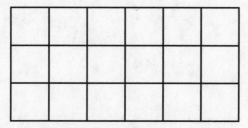

A rectangle partitioned into 3 rows with 6 squares in each row

Please return this Home Link to school tomorrow.

① Partition this rectangle into 2 rows with 3 same-size squares in each row.

How many squares cover the rectangle? _____

Practice

②   18
    + 21

③ 46 + 34 = _____

④   59
    + 37

Unit

Familiar Groups and Arrays

Family Note

In today's lesson your child solved number stories about equal groups and arrays. Equal groups are groups that all have the same number of objects. Arrays are rectangular arrangements of objects or symbols in rows and columns. Arrays show equal groups because each row in an array has the same number of objects, and each column has the same number of objects. You can find equal groups and arrays in many real-life objects and situations, such as those shown below. Your child can find the total number of objects efficiently by adding the number of objects in each group, row, or column. For example, to find how many dots are in 2 rows of 3 dots each on a die, you could find $3 + 3 = 6$. Or you could view the die as having 3 columns of 2 dots each and find $2 + 2 + 2 = 6$. To find how many fingers are on 2 hands with 5 fingers each, you could add $5 + 5 = 10$.

Please return this Home Link to school tomorrow.

Find the total number of objects in each picture. Then write a number model.

Example:

There are __6__ dots.

Number model:

$3 + 3 = 6$

①

There are _____ fingers in all.

Number model:

_____

②

There are _____ muffin cups.

Number model:

_____

③

There are _____ shoes in all.

Number model:

_____

Drawing Arrays

Family Note

In today's lesson your child used counters to show equal groups and arrays and then wrote number models to represent the counters. Encourage your child to use pennies or other small objects to help solve these problems.

Please return this Home Link to school tomorrow.

① Draw 4 equal groups with 3 in each group.

Number model:

_____

How many in all? _____

② Draw an array with 5 rows and 3 objects in each row.

Number model:

_____

How many in all? _____

③ Draw an array with 2 columns and 4 objects in each column.

Number model:

_____

How many in all? _____

Practice

Solve.

④
```
   23
+ 32
```

⑤ 63 + 28 = _____

⑥
```
   45
+ 49
```

Unit

Playing Array Concentration

Family Note

Today your child played a game called *Array Concentration* to practice finding the total number of objects in arrays and writing matching addition number models. In this game children match each array card with the number card that shows the total number of dots in the array. For example, the array card and the number card at the right "match."

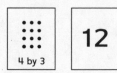

4 by 3   12

Please return this Home Link to school tomorrow.

Celia is playing *Array Concentration*. Her matches are shown below. Fill in the numbers on the number cards and write number models for the arrays.

Example:

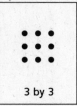

3 by 3   9

Number model:

$3 + 3 + 3 = 9$

①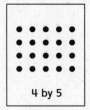

4 by 5

Number model:

_____

②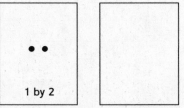

1 by 2

Number model:

_____

③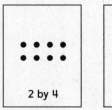

2 by 4

Number model:

_____

Writing a Shape Riddle

Family Note

In this lesson your child learned to recognize a 2-dimensional shape based on specific attributes, such as the following:

- number of angles
- number of sides
- number of pairs of parallel sides
- number of right angles

Using these attributes of 2-dimensional shapes, ask your child to write a shape riddle. *For example:* I am a shape that has 3 sides and 3 angles. I have no parallel sides. What shape am I? (The answer is "a triangle.") Your child can share the riddle with a family member or a friend.

Please return this Home Link to school tomorrow.

(1) Make up your own shape riddle. Give it to someone to solve.

_____

_____

_____

_____

_____

Equal Shares and Whole Number Operations

In Unit 9 children partition shapes into same-size parts, or equal shares. They practice using fraction vocabulary to name these equal shares and learn that equal shares do not necessarily have to be the same shape.

These equal shares are the same shape.

These equal shares are not the same shape.

Children also work with fractional units of length. They identify half-inches and quarter-inches on their rulers and measure objects to the nearest half-inch.

Later in the unit, children extend their work with place value to the thousands place and apply their understanding of place value to learn a new subtraction method called *expand-and-trade subtraction*. Children learn the expand-and-trade method by using expanded form to think about making trades.

Example: 45 − 27.

$$
\begin{array}{r}
45 \rightarrow \overset{30}{\cancel{40}} + \overset{15}{\cancel{5}} \\
-\ 27 \rightarrow \underline{20 + 7} \\
10 + 8 = 18
\end{array}
$$

Expand-and-trade subtraction will be reviewed in Grade 3. By the end of Grade 2, children are expected to subtract within 1,000 using any strategy or method.

In the final part of the unit, children review the values of coins and find coin combinations to pay for a variety of items using exact change. They use dimes and nickels as a context for finding multiples of 10 and 5 and also use doubling and doubles facts as a context for finding multiples of 2. These activities lay the foundation for multiplying by 2, 5, and 10 early in Grade 3.

Please keep this Family Letter for reference as your child works through Unit 9.

Vocabulary   Important terms in Unit 9:

one-half (1-half) A name for 1 out of 2 equal shares. The standard notation for one-half is $\frac{1}{2}$, but children do not use standard notations in Grade 2.

two-halves (2-halves) A name for the whole when it is divided into 2 equal shares. The standard notation for two-halves is $\frac{2}{2}$.

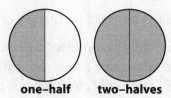

one–half   two–halves

one-fourth (1-fourth) A name for 1 out of 4 equal shares. The standard notation for one-fourth is $\frac{1}{4}$. Also called *one-quarter* or *1-quarter*.

four-fourths (4-fourths) A name for the whole when it is divided into 4 equal shares. The standard notation for four-fourths is $\frac{4}{4}$. Also called *four-quarters* or *4-quarters*.

one–fourth   four–fourths
or   or
one–quarter four–quarters

equal share Another name for equal parts. The result of dividing something into parts that are all the same size.

Home Links 9-1 and 9-2 provide more information about equal shares and the fraction language that appears in the definitions on this page.

one-third (1-third) A name for 1 out of 3 equal shares. The standard notation for one-third is $\frac{1}{3}$.

three-thirds (3-thirds) A name for the whole when it is divided into 3 equal shares. The standard notation for three-thirds is $\frac{3}{3}$.

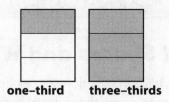

one–third   three–thirds

thousand cube In *Everyday Mathematics,* a base-10 block that measures 10 cm by 10 cm by 10 cm. A thousand cube consists of one thousand 1-centimeter cubes.

A thousand cube

expand-and-trade subtraction A subtraction algorithm in which expanded notation is used to facilitate place-value exchanges. Home Links 9-6 and 9-7 provide more information about expand-and-trade subtraction.

multiple The product of a certain number and any counting number. For example, the multiples of 2 are 2, 4, 6, 8, and so on (because those numbers are obtained by multiplying 2 by 1, 2, 3, 4, and so on, respectively). The multiples of 5 are 5, 10, 15, 20, and so on. And the multiples of 10 are 10, 20, 30, 40, and so on.

Do-Anytime Activities

To work with your child on Grade 2 concepts, try these interesting and rewarding activities:

1. Ask your child to divide food items or other objects into 2, 3, or 4 equal parts. For example, ask your child to fairly share a sandwich with a sibling or cut a piece of paper into four pieces that are the same size. Ask your child to name the parts of the object using language such as *one-half, 1-third,* or *1 out of 4 equal parts.* Then ask your child to name the whole object using language such as *whole, three-thirds,* or *4-fourths.*

2. Have your child measure the lengths of objects to the nearest inch and use the measurements to compare the objects. When your child is comfortable measuring to the nearest inch, have him or her measure the same objects to the nearest half-inch.

3. Pose subtraction problems involving 2-digit numbers and ask your child to explain his or her strategy for solving them.

4. Point to an item in a store or an ad and have your child tell you what coins and bills he or she could use to pay for the item with exact change.

Building Skills through Games

In Unit 9 your child will play the following games to practice his or her mathematical skills.

Array Concentration

Players arrange a set of *Array Concentration* Number Cards and Array Cards facedown in front of them. A player flips over one of each type of card. If the cards "match"—that is, if the number on the number card equals the total number of dots in the array—the player takes the cards and takes another turn.

Shape Capture

Players have a set of Shape Cards spread out in front of them. One at a time players draw an Attribute Card and "capture" all the shapes that have that attribute. The player who captures the most shapes wins.

Beat the Calculator

One player is the Caller, who names two 1-digit numbers. Another player is the Brain, who adds the two numbers mentally. A third player is the Calculator, who adds the numbers with a calculator. The Brain tries to find the sum faster than the Calculator.

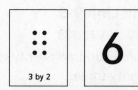

These cards match because there are 6 dots in the array.

Hit the Target

Players choose a 2-digit multiple of 10 (such as 10, 20, or 30) as a target number. One player chooses a starting number less than or greater than the target number, which the second player enters into a calculator. The second player tries to change it to the target number by adding or subtracting on the calculator.

As You Help Your Child with Homework

When your child brings home assignments, you may want to go over the instructions together, clarifying them. The answers listed below will guide you through the Unit 9 Home Links.

Home Link 9-1

1. one-half; 1-half; 1 out of 2 equal parts; 2-halves; two-halves; 2 out of 2 equal parts

2. 1 out of 4 equal parts; 1-fourth; one-quarter; whole; four-fourths; 4 out of 4 equal parts

Home Link 9-2

1. Sample answer: 1 out of 2 equal parts; 2 out of 2 equal parts

2. Sample answer: 1-third; three-thirds

Home Link 9-3

1. Sample answer:

2. Sample answer: Cut the rectangle out and fold it along the lines to see if the parts are the same size.

3. Sample answers: 1-fourth; one-quarter

4. Sample answers: four out of four equal shares; 4-fourths

5. 107   **6.** 47   **7.** 82

Home Link 9-4

1. About 2 inches

2. Possible answers: 3 and one-half; 3 and 1-half

3. About 2 inches   **4.–7.** Answers vary.

Home Link 9-5

1. 329   **2.** 183

3. Three hundred twenty-nine; one hundred eighty-three

4. $400 + 90 + 1$ $400 + 70 + 1$ $491 > 471$

5. <   **6.** >   **7.** 158   **8.** 26   **9.** 102

Home Link 9-6

1. Sample estimates: $50 - 30 = 20$; $60 - 35 = 25$
   Sample sketch:

      Answer: 19

2. Sample estimate: $60 - 30 = 30$
   Sample sketch:

   Answer: 36

Home Link 9-7

1. Sample estimate: $60 - 40 = 20$

$$
\begin{array}{ccc}
 & & 40 \quad 15 \\
55 & \rightarrow & \cancel{50} + \cancel{5} \\
-\ 37 & \rightarrow & 30 + 7 \\
\hline
 & & 10 + 8 = 18
\end{array}
$$

2. Sample estimate: $80 - 30 = 50$

$$
\begin{array}{ccc}
 & & 70 \quad 11 \\
81 & \rightarrow & \cancel{80} + \cancel{1} \\
-\ 28 & \rightarrow & 20 + 8 \\
\hline
 & & 50 + 3 = 53
\end{array}
$$

Home Link 9-8

1. Possible answers: 10¢ or $0.10; 50¢ or $0.50; 100¢ or $1.00; 250¢ or $2.50

2. Answers vary.

Home Link 9-9

1–2. Sample explanations given.

1. No. 59¢ is almost 60¢, and 49¢ is almost 50¢. 60¢ + 50¢ is more than $1.

2. No. $30 + 10 = 40$ and 2 and 8 make another 10, so the total for the radio and headphones is $50. I couldn't buy the calculator, too.

3. 38   **4.** 91   **5.** 25

Home Link 9-10

1. 14 fingers; $7 + 7 = 14$

2. 4 shells; $4 + 4 = 8$

3. 58   **4.** 130   **5.** 25

Home Link 9-11

1. 10 cents, 10, 10; 30 cents, 30, 30

2. 40 cents, 40, 40; 70 cents, 70, 70

3. 80 cents, 80, 80; 40 cents, 40, 40

4. 140   **5.** 43   **6.** 175

Equal Shares

Family Note

In this lesson your child divided shapes into 2, 3, or 4 equal parts and used words to first name 1 equal part and then to name all of the equal parts together. For example, the square at the right is divided into 3 equal parts.

Each of the parts can be named *one-third, 1-third,* or *1 out of 3 equal parts*. All of the parts together can be named *three-thirds, 3-thirds,* or *3 out of 3 equal parts*. Although your child may have had experience with standard notation for fractions ($\frac{1}{3}$, $\frac{3}{3}$, and so on), the formal introduction of standard notation occurs in third grade.

Please return this Home Link to school tomorrow.

① Divide this square into 2 equal parts.

Circle names for 1 part.

one-half   1-half   2 out of   1 out of
                              3 equal parts   2 equal parts

Circle names for all of the parts.

1 out of   2-halves   two-halves   2 out of
3 equal parts   2 equal parts

② Divide this square into 4 equal parts.

Circle names for 1 part.

1 out of   1-fourth   1 out of   one-quarter
4 equal parts   3 equal parts

Circle names for all of the parts.

whole   four-fourths   one-quarter   4 out of
                                              4 equal parts

Fraction Names

Family Note

In today's lesson your child used pattern blocks to divide shapes and then used fraction words to name the equal parts. As you work through the activity below, guide your child to use names from the Fraction Names Word Box. Discuss the names of the parts of each shape.

Please return this Home Link to School tomorrow.

MRB
132–133

Fraction Names Word Box	
Names for 1 Part	**Names for All of the Parts**
one-half, 1-half, 1 out of 2 equal parts	two-halves, 2-halves, 2 out of 2 equal parts, whole
one-third, 1-third, 1 out of 3 equal parts	three-thirds, 3-thirds, 3 out of 3 equal parts, whole

① Write the name for 1 part.

_____

Write the name for all of the parts.

_____

② Write the name for 1 part.

_____

Write the name for all of the parts.

_____

Naming Equal Shares

Family Note

In this lesson we continued making and naming equal shares of rectangles and circles. Your child showed and described how to share 3 muffins equally between 2 children, and 5 muffins equally among 4 children. By solving and discussing problems like these, your child will learn appropriate fraction vocabulary, such as 1 out of 2 equal shares, one-half, 1-third, one-quarter, 1-fourth, and one out of four equal shares. Practice making and naming fractional amounts will continue to the end of the year and will lead to a great deal of work with fractions in *Third Grade Everyday Mathematics*.

Please return this Home Link to school tomorrow.

① Divide the rectangle into 4 equal parts.

② How could you test that the parts are equal?

_____

_____

③ Use words to name one of the parts in at least two ways.

_____   _____

④ Use words to name all of the parts together.

_____

Practice

Unit

⑤
```
   73
 + 34
```

⑥
```
   90
 − 43
```

⑦
```
   46
 + 36
```

Measuring Lengths

Family Note

Today your child measured life-size pictures of objects to the nearest inch and half-inch. Because standard notation for fractions ($\frac{1}{2}$, $\frac{1}{3}$) has not been introduced, we recorded half-inch measures with fraction words such as *one-half* or *1-half*. Pick a few objects and take turns with your child measuring each one to the nearest inch and half-inch. Compare your measurements to ensure they are the same.

Please return this Home Link to school tomorrow.

Cut out the 6-inch ruler below. Use it to measure these line segments to the nearest inch.

① _____ About _____ inches

Measure these line segments to the nearest half-inch.

② _____

About _____ inches

③ _____ About _____ inches

Measure some objects at home to the nearest half-inch. List the objects and their measurements below.

④ _____

⑤ _____

⑥ _____

⑦ _____

✂

0	1	2	3	4	5	6

inches (in.)

Place Value

Family Note

In this lesson your child reviewed place value and how it is used to determine the value of digits in numbers. For example, the 5 in 503 is worth 5 hundreds, or 500, because it is in the hundreds place. The 5 in 258 is worth 5 tens, or 50, because it is in the tens place.

Your child also used place value to compare numbers. For example, to compare 571 and 528, your child might think, "Both numbers have 5 hundreds. But 571 has 7 tens and 528 has only 2 tens. So 571 is the larger number."

Please return this Home Link to school tomorrow.

In Problems 1–2, write the numbers shown by the base-10 blocks.

①  _____

②  _____

③ Read the numbers in Problems 1–2 aloud to someone at home.

④ Write each number in expanded form. Then write < or > in the box to compare the two numbers.

491 = _____

471 = _____

491 ☐ 471

Write <, >, or =.

⑤ 295 ☐ 298

⑥ 387 ☐ 378

Practice

Add or subtract.

Unit ☐

⑦ 93 + 65 = _____

⑧ 80 − 54 = _____

⑨ 76 + 26 = _____

Making Trades to Subtract

Family Note

In this lesson your child learned about subtracting multidigit numbers using base-10 blocks. Your child also used ballpark estimates to check whether answers made sense. When using base-10 blocks to subtract, children first check if they need to make any trades. All trades are made before any subtraction is done. Trading first allows children to concentrate on one thing at a time.

Example: 62 − 36 = ?
- Make a ballpark estimate: 62 is close to 60, and 36 is close to 40, so one estimate is 60 − 40 = 20.
- Sketch 62 using base-10 shorthand:

- Are there enough longs and cubes to remove 3 longs and 6 cubes (36)? No, so you need to trade.
- Trade 1 long for 10 cubes:
- Does the sketch still show 62? Yes.

- Can we remove the blocks now to help subtract 36? Yes.
- Remove them.

- Count the longs and cubes that are left. The answer is 26.
- Check to see whether the answer makes sense. The ballpark estimate of 20 is close to the answer of 26, so 26 is a reasonable answer.

Please return this Home Link to school tomorrow or as requested by the teacher.

---

① 53
  − 34

② 64
  − 28

Ballpark estimate:

_____

Sketch 53 using base-10 shorthand. Solve the problem. Show your work.

Ballpark estimate:

_____

Sketch 64 using base-10 shorthand. Solve the problem. Show your work.

Answer: _____

Answer: _____

Explain to someone how you know your answers make sense.

Expand-and-Trade Subtraction

Family Note

In this lesson your child subtracted multidigit numbers using expand-and-trade subtraction. Instead of using base-10 blocks, your child used expanded form to think about making trades. Your child continued to use ballpark estimates to check whether answers made sense.

Example: $62 - 36 = ?$

- Write a number sentence to show a ballpark estimate: $60 - 40 = 20$.

- Write each number in expanded form.

$$62 \rightarrow 60 + 2$$
$$-36 \rightarrow 30 + 6$$

- Look at the 10s and 1s. Can you subtract without making trades? No; so trade 1 ten for 10 ones. Cross out 60 (6 tens) and replace it with 50 (5 tens). Cross out 2 (2 ones) and replace it with 12 (12 ones). Then subtract.

$$62 \rightarrow \overset{50}{\cancel{60}} + \overset{12}{\cancel{2}}$$
$$-36 \rightarrow 30 + 6$$
$$20 + 6 = 26$$

Add the tens and ones to find the total: $20 + 6 = 26$. So $62 - 36 = 26$.

- Compare your answer to your estimate: 20 is close to 26, so 26 is a reasonable answer.

Please return this Home Link to school tomorrow or as requested by the teacher.

Use expand-and-trade subtraction to solve. Use a ballpark estimate to check your answer.

MRB
81,
87–89

① $55 - 37 = ?$

Ballpark estimate:

_____

Solution:

$55 - 37 = $ _____

② $81 - 28 = ?$

Ballpark estimate:

_____

Solution:

$81 - 28 = $ _____

Copyright © McGraw-Hill Education. Permission is granted to reproduce for classroom use.

Coin Combinations

Family Note

In today's lesson children practiced writing money amounts in cents notation and dollar-and-cents notation. In Problem 1, for example, your child might write the value of 10 pennies as 10¢ or $0.10. Your child also showed two different ways to pay for a single item. For example, your child might have shown 62¢ with 2 quarters, 1 dime, and 2 pennies or with 4 dimes, 4 nickels, and 2 pennies. For Problem 2, help your child find items costing less than 99¢ in newspaper or magazine ads and find different combinations of coins to pay for the items.

Please return this Home Link to school tomorrow.

① Pretend that you have 10 of each kind of coin. How much money would you have?

Fill in the blanks.

10 pennies = _____

10 nickels = _____

10 dimes = _____

10 quarters = _____

② Find two ads in a newspaper or magazine for items that cost less than 99¢ each.

- Ask for permission to cut out the ads.

- Cut them out and paste or tape them onto the back of this page.

- Draw coins to show two different ways to pay for each item with exact change.

(If you can't find ads, draw pictures of items and prices on the back of this page.)

Estimating Total Cost

Family Note

In this lesson we worked on a problem in which your child pretended to be at a store and needed to estimate the total cost of selected items using mental math. When you are in a store together, choose two or three items and ask your child to try to estimate the total cost without using pencil and paper. Encourage the use of "close-but-easier" numbers for each item to make it easier to find the total cost using mental math.

Please return this Home Link to school tomorrow.

For each problem, pretend you are at a store and do not have a calculator or pencil and paper.

MRB
79

① You have $1. You want to buy a toy for 59¢ and an apple for 49¢. Do you have enough money? Tell why or why not.

② You have $50. You want to buy a radio for $32, headphones for $18, and a calculator for $6. Do you have enough money? Tell why or why not.

Practice

Unit

Add or subtract.

③   67
   − 29

④ 35 + 56 = _____

⑤   71
   − 46

Two Equal Groups

Family Note

In this lesson your child solved problems involving 2 equal groups. In some of the problems, your child needed to find the total number of objects in 2 equal groups.

Example: There are 2 packages of water bottles. Each package has 6 bottles. How many bottles are there in all?
Answer: 12 bottles

Your child can use doubles facts to help solve these problems. For the above problem, your child might think "What is the double of 6? The double of 6 is 12 because $6 + 6 = 12$."

In other problems, your child needed to share items equally into 2 groups.

Example: You have 10 dishes that you want to put in 2 equal piles. How many dishes should you put in each pile?
Answer: 5 dishes

Your child can also use doubles facts to help solve these problems. For the above problem, your child might think "Which doubles fact has 10 as the sum? It's $5 + 5 = 10$, so there are 5 in each pile."

Please return this Home Link to school tomorrow.

Solve each problem and write a number model.

MRB
32–33

① A space alien has 2 hands with 7 fingers on each hand. How many fingers does the space alien have in all?

Answer: _____ fingers
Addition number model:

_____

② You have 8 shells to give to 2 friends. You give the same number to each friend. How many shells does each get?

Answer: _____ shells
Addition number model:

_____

Practice

Add or subtract.

Unit

③ 77
   −19

④ $47 + 83 =$ _____

⑤ 51
   − 26

5s and 10s

Family Note

In this lesson your child solved problems involving multiples of 10 and 5. A multiple of 5 is the answer to a multiplication problem involving 5 and any counting number. For example, 20 is a multiple of 5 because $5 \times 4 = 20$. The number 20 is also a multiple of 10 because $10 \times 2 = 20$.

The multiples of a number are also the skip counts of that number.

Multiples of 5: 5, 10, 15, 20, . . .   **Multiples of 10:** 10, 20, 30, 40, . . .

Dimes and nickels were used as a context for finding multiples of 5 and 10. Your child can solve the problems below by skip counting.

Please return this Home Link to school tomorrow.

MRB
32–33

① 2 nickels = _____ cents   2 [5s] is _____ $2 \times 5 =$ _____

    6 nickels = _____ cents   6 [5s] is _____ $6 \times 5 =$ _____

② 4 dimes = _____ cents   4 [10s] is _____ $4 \times 10 =$ _____

    7 dimes = _____ cents   7 [10s] is _____ $7 \times 10 =$ _____

③ 8 dimes = _____ cents   8 [10s] is _____ $8 \times 10 =$ _____

    8 nickels = _____ cents   8 [5s] is _____ $8 \times 5 =$ _____

Practice

Add or subtract.

Unit

④   46
   + 94
   _____

⑤ $92 - 49 =$ _____

⑥   99
   + 76
   _____

Congratulations!

By completing *Second Grade Everyday Mathematics,* your child has accomplished a great deal. Thank you for your support!

This Family Letter is provided as a resource for you to use throughout your child's vacation. It includes an extended list of Do-Anytime Activities, directions for games that can be played at home, and a sneak preview of what your child will be learning in *Third Grade Everyday Mathematics.* Enjoy your vacation!

Do-Anytime Activities

Mathematics concepts are more meaningful and easier to understand when they are rooted in real-life situations. To help your child review some of the concepts he or she has learned in second grade, we suggest the following activities for you and your child to do together over vacation. Doing so will help your child maintain and build on the skills learned this year and help prepare him or her for *Third Grade Everyday Mathematics.*

1. Pose addition and subtraction number stories about everyday life. For example, ask your child to count the number of grapes he or she has and then ask: *How many will you have if you eat 6 of them? How many will you have if you eat 2 of them and then I eat 3 more?* Here's another example: *If you have 1 quarter, 3 dimes, and 2 nickels, how many cents do you have?*

2. Review and practice addition and subtraction facts. Your child can use Fact Triangle cards to practice or play *Addition Top-It* or *Subtraction Top-It* as described on the second page of this letter.

3. Select everyday objects and have your child estimate their lengths and then measure to check the estimates. Your child could also measure objects to determine how much longer one thing is compared with another.

4. Ask your child to tell you the time to the nearest 5 minutes. Encourage your child to specify whether it is A.M. or P.M.

5. Encourage your child to identify and describe geometric shapes that can be seen in the world. For example: *I see rectangles in that bookcase. They all have 4 right angles.* You can also play *I Spy* to practice identifying and describing shapes. For example: *I spy a shape with 5 sides. All of the sides are the same length.*

6. Ask your child to share food items or other objects fairly with 1, 2, or 3 other people by dividing them into equal shares.

7. Count on or back by 10s and 100s from any given number.

Building Skills Through Games

This section describes games that can be played at home. The number cards used in some games can be made from 3"-by-5" index cards or from a regular playing-card deck. (Use jacks for zeros and write the numbers 11 through 20 on the four queens, four kings, and two jokers.)

Addition Top-It

Materials	4 cards for each of the numbers 0–10
Players	2 or more
Skill	Adding two numbers
Object of the game	To have the most cards

Directions

Shuffle the cards and place them facedown in a pile. Each player turns up a pair of cards from the deck and says the sum of the numbers. The player with the greater sum takes all the cards from that round. Players continue turning up cards and saying the sums until there are no more cards left in the draw pile. The player with the most cards at the end of the game wins.

Variation: *Subtraction Top-It*

Add cards for the numbers 11–20 to the *Addition Top-It* deck. Each player turns up a pair of cards from the deck and says the difference between the two numbers. The player with the greater difference takes all the cards from that round.

Salute!

Materials	4 cards for each of the numbers 0–10
Players	3
Skill	Finding missing addends
Object of the game	To have the most cards

Directions

Shuffle the cards and place them facedown in a pile. One person is the Dealer and gives the two Players one card each. Without looking at the numbers, the Players place the cards on their foreheads facing out, so everyone can see the numbers. The Dealer, who sees both numbers, says the sum of the two cards. The others use the sum and the number on the other card to figure out the number on their foreheads. The Player that finds his or her number first takes both cards. Players rotate roles, with someone new taking over as Dealer in each round. Play continues until everyone has been Dealer five times. The one with the most cards at the end is the winner.

Sample round:

Tom is the Dealer. He gives Raul a 5 and Cheri a 7. Tom looks at both cards and says, "The sum is 12." Raul can see Cheri's 7 and thinks, "What plus 7 is 12?" Raul says, "My number is 5." Because he figures out his number faster than Cheri figures out hers, Raul takes both cards.

Name That Number

Materials	4 cards for each of the numbers 0–10
	1 card for each of the numbers 11–20
Players	2 or 3
Skill	Adding or subtracting numbers to reach a target number
Object of the game	To have the most cards

Directions

Shuffle the cards and place them facedown in a pile. Turn the top five cards faceup and place them in a row. Turn over the next card and place it faceup by the pile. This is the target number.

Players take turns trying to name the target number by adding or subtracting the numbers on two or more of the five cards that are faceup. Cards may be used only once for each turn. When a player is unable to name the target number using the faceup cards, his or her turn is over. The target is replaced with a card drawn from the top of the deck.

When players are able to name the target number, they collect the cards they used to name it along with the target-number card. Replacement cards for the five faceup cards are drawn from the deck. The next card from the top of the deck is the new target number.

Play continues until there are not enough cards left in the deck to replace the faceup cards. The player who has collected the most cards wins.

Sample turn:

Mae's turn:

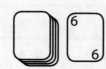

The target number is 6. Mae names it with 12 − 4 − 2. She could also have used 4 + 2 or 8 − 2. Mae takes the 12, 4, 2, and 6 cards. She replaces them by drawing cards from the deck as well as a new target number. Now it is Mike's turn.

Hit the Target

Materials	calculator
	record sheet (see example below)

Target number: 30

Starting Number	Change	Result	Change	Result	Change	Result
17	+23	40	−10	30		

Players	2
Skill	Finding differences between 2-digit numbers and multiples of 10
Object of the game	To reach the target number.

Directions

Players agree on a multiple of 10 (10, 20, 30, 40, and so on) as a target number and write it on the record sheet. Player 1 names a starting number that is less than or greater than the target number and records it on the record sheet. Player 2 enters the starting number on a calculator and tries to hit the target number by adding or subtracting a number to it. Player 2 continues adding and subtracting until he or she reaches the target number, recording the change and results on the record sheet. Then players switch roles: Player 2 chooses a starting number and Player 1 tries to change the starting number to the target number by adding and subtracting on the calculator. The player who reaches the target number in fewer tries wins the round.

Sample turn:

Kylie and Aiden agree on 30 as the target number. Kylie chooses 17 as the starting number. Aiden tries to change 17 to 30 by adding 23 but gets a result of 40. He subtracts 10, hitting the target in two tries. His record sheet looks like the one shown on page 284.

Fact Power

Another way addition and subtraction facts can be practiced is by using the Addition/Subtraction Facts Table shown below. The table can also be used to keep a record of facts that have been learned. For example, your child might color the squares for the sums that he or she knows from memory.

+, −	0	1	2	3	4	5	6	7	8	9
0	0	1	2	3	4	5	6	7	8	9
1	1	2	3	4	5	6	7	8	9	10
2	2	3	4	5	6	7	8	9	10	11
3	3	4	5	6	7	8	9	10	11	12
4	4	5	6	7	8	9	10	11	12	13
5	5	6	7	8	9	10	11	12	13	14
6	6	7	8	9	10	11	12	13	14	15
7	7	8	9	10	11	12	13	14	15	16
8	8	9	10	11	12	13	14	15	16	17
9	9	10	11	12	13	14	15	16	17	18

Looking Ahead:
Third Grade Everyday Mathematics

Next year your child will . . .

- Learn multiplication facts.

- Explore the relationship between multiplication and division.

- Write number models for addition, subtraction, multiplication, and division number stories.

- Further explore addition and subtraction of 2- and 3-digit numbers.

- Continue partitioning figures and number lines to build an understanding of fractions.

- Tell time to the nearest minute.

- Measure length to the nearest quarter inch.

- Find perimeters and areas of rectangles.

- Further explore the attributes of shapes.

Again, thank you for your support this year. Have fun continuing your child's mathematical adventures throughout the vacation.